THE BIGGEST BANGS

The Biggest Bangs

The Mystery of Gamma-ray Bursts, the Most Violent Explosions in the Universe

Jonathan I. Katz

OXFORD
UNIVERSITY PRESS

2002

OXFORD

UNIVERSITY PRESS

Oxford   New York
Auckland   Bangkok   Buenos Aires   Cape Town
Chennai   Dar es Salaam   Delhi   Hong Kong   Istanbul   Karachi   Kolkata
Kuala Lumpur   Madrid   Melbourne   Mexico City   Mumbai   Nairobi
São Paulo   Shanghai   Singapore   Taipei   Tokyo   Toronto

and an associated company in Berlin

Published by Oxford University Press, Inc.
198 Madison Avenue, New York, New York 10016

Oxford is a registered trademark of Oxford University Press

Library of Congress Cataloging-in-Publication Data
Katz, Jonathan I.
The biggest bangs: the mystery of gamma-ray bursts, the most violent explosions
in the universe / by Jonathan I. Katz.
p. cm.   Includes bibliographical references and index.
ISBN 0-19-514570-4
1. Gamma-ray bursts.   I. Title.
QB471.7.B85 K38 2002
522'.6862—dc21   2001036545

9 8 7 6 5 4 3 2 1

Printed in the United States of America
on acid-free paper

This book is dedicated by the author to his children,
the joy of his manhood,
the pride of his maturity,
the consolation of his age,
and the posterity of his ancestors.

Contents

Preface ix

Introduction 1

1. Vela 3

2. Detectors 12

3. Where Are They? 21

4. What Are They? 29

5. Compactness 40

6. The Large Magellanic Cloud 50

7. False Lines 60

8. False Light 70

9. The Copernican Dilemma 82

10. Soft Gamma Repeaters 94

11. BATSE 106

12. The Great Debate 116

13. The Theorists' Turn 126

14. Afterglows 139

15. A Supernova Connection? 152

16. The Holy Grail 162

17. The End of the Beginning 178

Afterword 181

*Appendix. Did a Gamma-ray Burst Kill the Dinosaurs?
Will a Burst Kill Us?* 185

Glossary 189

Sources 197

Index 209

Preface

During the decade 1962–73 a series of remarkable discoveries—quasars, X-ray stars, the cosmic microwave background radiation, pulsars, and finally gamma-ray bursts—turned astronomy from a sleepy backwater of science into one of its boom towns. Of these discoveries, gamma-ray bursts were the most enigmatic. In contrast to quasars, X-ray stars, and pulsars, whose origins and mechanisms were understood, at least in outline, within a year of their discovery, and in contrast to the microwave background radiation, which had actually been predicted from cosmological theory decades earlier, gamma-ray bursts remained mysterious for a quarter of a century. Many of the more ambitious theoretical astrophysicists tried their hands at explaining them. There were no Einsteins, Watsons, or Cricks; it seemed that every wrong path and dead end was explored. Our present understanding emerged gradually, bit by bit. Fundamental questions are still unanswered. Even today, we understand gamma-ray bursts no better than quasars were understood in 1964, not long after their discovery.

This is very different from the traditional picture of scientific progress, in which a problem is solved by a single blinding insight or spectacular discovery. The reason is that gamma-ray bursts are a very complicated phenomenon, in which each part, difficult to understand in itself, depends on the results of the preceding parts. Their solution requires solving several different but connected puzzles. Many apparently illuminating data turned out to be misleading, and many promising ideas turned out to be dead ends. The purpose of this book is to tell the story of how many scientists, over many years, stumbled toward the solution of this problem.

During the heyday of theoretical astrophysics it was hoped that study of new astronomical phenomena would lead to the discovery of new physical laws. This hope was not fulfilled, and there is no evidence that fundamentally new physical phenomena lie behind the discoveries of astronomers. Instead, familiar phenomena come together in novel ways. In fact, the chief problem of theoretical astrophysics is the problem of initial conditions: where were the matter and energy, and what were they doing, before the action began. Some think that understanding these initial conditions and new combinations is less important, or less exciting, than the discovery of new laws of nature, but this is a subjective and aesthetic judgment, not a self-evident truth.

In recent decades some physicists have turned the search for new fundamental laws of nature into an exercise in mathematics and speculation, sterile because the domain of applicability of these laws lies far beyond any foreseeable experiment. Rather, new laws must be found in more complicated, but observable, phenomena. These laws will not take the form of new fundamental interactions or elementary particles, but will concern the construction of complex phenomena from simpler ingredients. Most of the pieces necessary to understand gamma-ray bursts were available more than a half-century ago, but no one could have predicted how they would combine to make a burst. Nature is much cleverer and more inventive than the human mind.

Working scientists use the units that are most natural in their fields. In astronomy large distances are usually measured in parsecs (the parsec is defined as 206,265 times the mean distance between Earth and the Sun). Because this unit is unfamiliar to most laymen, I have instead used light-years, the distance light travels in a year. The nearest star (other than the Sun) is about four light-years (1.3 parsecs) away. Smaller distances are usually measured in centimeters. Other units are used as clear communication requires. Most physical scientists do not limit themselves to SI units (a particular version of metric units) because these are often inconvenient or confusing. Units should be our tools, not our masters.

Finally, although this book is written for the layman it has been necessary to make one concession to the language of science, the use of "scientific notation": astronomy involves very big and very small numbers, which would be cumbersome to write out. In scientific notation 10^{52} means 1 followed by 52 zeros (!), while 10^{-12} means a decimal point followed by eleven zeros, and then a 1 in the 12th place.

In the course of writing this book I have been aided by many people. I thank C. W. Akerlof, W. Alvarez, J. Apt, S. D. Barthelmy, W. R. Binns, P. Boynton, J. H. Buckley, D. E. Casperson, A. Dar, J. E. Felten, G. J. Fishman, D. A. Frail, T. Galama, N. Gehrels, J. Goodman, P. Groot, M. M. Halloran, P. Hink, M. H. Israel, P. C. Joss, S. M. Kahn, R. W. Klebesadel, C. Kouveliotou, B. C. Lee, D. R. Lorimer, R. A. Muller, W. S. Paciesas, H.-S. Park, P. J. E. Peebles, T. Piran, W. Press, T. A. Prince, M. J. Rees, G. R. Ricker, R. Roberts, J. M. Ryan, K. C. Sahu, B. E. Schaefer, K. Slavis, D. A. Smith, "Deep Space," G. Tarle, B. Teegarden, V. Trimble, J. Trümper, V. V. Usov, J. Wren, M. Zimmermann, and A. Żytkow, who have clarified and explained many points, but who are not responsible for any remaining errors or omissions. I also extend thanks and an apology to all those not mentioned here who worked on the gamma-ray burst mystery, which has fascinated us for many years. Special thanks are due my wife.

September 2001   J. I. K.
St. Louis, Missouri

THE BIGGEST BANGS

Introduction

Gamma-ray bursts are the most violent events since the birth of the universe. They are perhaps ten times as energetic as the most energetic supernovas, the explosions that destroy massive stars and make neutron stars and black holes. More remarkable, the energy in a gamma-ray burst is concentrated into a small amount of mass moving at 99.999% of the speed of light. It has even been suggested that the most energetic cosmic rays—single elementary particles, each with enough energy to lift a 150-pound man an inch off the floor—are produced in gamma-ray bursts. At their peak, gamma-ray bursts have been observed to be, by far, the brightest things in the universe, about one hundred thousand times brighter than an entire galaxy. A gamma-ray burst at the distance of the nearest star would, for a few seconds, outshine the Sun nearly a million-fold and would incinerate half the Earth.

Gamma-ray bursts were completely unexpected. In 1967 the Vela satellites, designed and built during the Cold War to monitor the Nuclear Test Ban Treaty outlawing atomic bomb tests in space, first observed a gamma-ray burst. Bursts soon acquired a reputation as the toughest nut to crack in astronomy. A generation of scientists tried themselves against this problem and were found wanting. Innumerable theoretical balloons were floated and soon sank. Elaborate observational efforts produced either no data or data that did not answer the important questions. Only a quarter of a century after their discovery did we come to any understanding of gamma-ray bursts, and it remains incomplete.

Astronomers find gamma-ray bursts hard to study because they are rare, ephemeral, and completely unpredictable. The longest last a few minutes, and some only a small fraction of a second. This is extraordinary in astronomy, where most phenomena continue (almost) forever. The sky appears the same, night after night. Even other transient events, such as solar flares, X-ray bursts, and the many types of variable and exploding stars, do not disappear completely, but leave behind a steady source of radiation that may be examined at leisure.

There are two great obstacles to the study of gamma-ray bursts. The first is that after their brief glory they vanish, almost without a trace. There is no remnant that can be studied in its quiescent state. Scientific progress usually begins with a serendipitous discovery, but its exploitation requires returning to it with ever more powerful experimental and observational tools. This is rather difficult if the object disappears in seconds.

The second obstacle arises from the difficulty of localizing a source of gamma rays on the sky. The instruments that detect gamma rays do not determine their arrival directions well or at all. Astronomers need precise directions to tell them where to point their telescopes. Gamma-ray detectors don't provide these.

In the early 1990s clever experimenters achieved a breakthrough with careful statistical studies of large numbers of poorly localized gamma-ray bursts. This avoided the obstacles to studying individual bursts. Several years later astronomers finally succeeded in precisely localizing a few gamma-ray bursts and studying their faint afterglows with powerful optical and radio telescopes. They turned out to be among the most distant objects in the universe, and these great distances implied that they were enormously powerful.

This book is about the lengthy and difficult process by which we have come to some understanding of gamma-ray bursts. There are no human heroes in this story, and no scientists covered themselves with glory; the glory is in the wonders of nature and the heavenly creations of God.

1

Vela

On December 7, 1941, the "date which will live in infamy," a surprise Japanese attack on Pearl Harbor, Hawaii, sank much of the U.S. Navy's Pacific Fleet, shocked Americans from their naïve isolationism, and carved a permanent mark in the national character: never again would the United States permit itself to be caught by surprise.

When the Cold War began, the United States invested heavily in technical means of warning of surprise attack. A Distant Early Warning system of radar stations was built across the Arctic, from Alaska through Canada to Greenland, linked to a central command post deep under Cheyenne Mountain, in Colorado. Defense Support Program satellites used infrared sensors to detect missile launches. Additional sensors searched for the light and radio waves produced by nuclear explosions. An extensive photographic reconnaissance system, first using aircraft and then satellites, watched the territory of the Soviet Union and other potential adversaries.

The Vela satellites were born of the "never be surprised" philosophy, but were not meant to warn of an actual attack. Rather, their mission was to monitor the Nuclear Test Ban Treaty, signed in 1963. This treaty forbade nuclear explosions in the atmosphere, underwater, and in space. Its purpose was partly symbolic, to satisfy a public desire for a visible sign that something was being done about the danger of nuclear weapons, and partly practical, to reduce the genuine hazard of radioactive fallout from nuclear explosions in the atmosphere. The practical goal was achieved: the signatory nations ceased atmospheric testing and the level of fallout and radioactive contamination rapidly decreased. Eventually, all nuclear powers signed. The symbolic goal was, of course, only symbolic; nuclear weapons development continued unabated, using test explosions performed underground, until the maturing technology and the end of the Cold War brought it to a logical close, at least in the major nuclear powers.

In space a nuclear explosion does not produce much visible light or heat, and no sound, but it does make abundant X rays and gamma rays, their more energetic cousins. These radiations are absorbed by Earth's atmosphere and, if produced in space, can only be detected by sensors in space. The Vela satellites were launched, beginning in 1963, into orbits approximately 65,000 miles above Earth's atmosphere, about a quarter as far from us as the Moon. They are much higher than most Earth satellites, which barely skim the atmosphere, less than 500 miles above Earth's surface. Vela orbits are nearly three times higher than the orbits of geosynchronous communications satellites, which are placed high enough that they will revolve once each day, keeping them directly over the same point on the equator at all times. Their high orbits gave the Vela satellites orbital periods of 4¼ days.

It requires more rocket fuel to launch a satellite into a high Vela orbit than into a lower orbit, but these high orbits were chosen for good reasons. First, the Vela satellites would generally be well off to the side of a line from Earth to the Moon. This meant that they could detect nuclear explosions behind the Moon, just as we step to the side to look around and behind an obstacle. The American concern with not being surprised extended even to explosions behind the Moon. Second, these orbits are above the Van Allen radiation belts of Earth, reducing the noise in the sensors. Third, by comparing the times of arrival of a brief gamma-ray signal at multiple satellites it would be possible to determine where it was produced. There were usually four Vela satellites operational at any time, enough to localize the source of a flash of gamma rays to a small compact region in space. If the data were exact, this region would be a single point, but all measured data have some uncertainty and inaccuracy. The more widely separated the satellites, the more accurate this localization would be.

On July 2, 1967, the Vela satellites recorded an unexpected burst of gamma rays. Examination of the data showed that it was not a nuclear weapons test because a bomb produces a characteristic gamma-ray signal. This consists of a very brief and intense burst, less than millionth of a second long, from nuclear processes in the explosion itself, followed by a leveling out and then a gradual fading as unstable nuclei decay. The observed signal was not like this at all. It lacked the intense initial flash, and had two distinct peaks rather than a steady fading. The Vela satellites had detected natural phenomena before, such as magnetic storms that occur where the solar wind rubs against Earth's magnetic field, and the new observation was never mistaken for a clandestine Soviet weapons test. The team of scientists at Los Alamos analyzing data from the Vela satellites filed it away as a mystery for later investigation. It was not urgent.   ·

The Vela system underwent regular upgrades, replacing older satellites with new, better ones. Beginning in 1969, these improved satellites began detecting gamma-ray bursts more often. In three years sixteen were observed. The Los Alamos team had not thought of themselves as astronomers, and their leader, Ray Klebesadel, had been trained as an electrical engineer, but they found themselves making one of the most remarkable and enigmatic astronomical discoveries ever. In 1973 they published their results.

The data from one burst are shown in Figure 1-1, taken from the discovery paper, the raw material of scientific history. Data from three Vela satellites were shown to convince the skeptical reader (the astronomer of 1973, who had not imagined that the universe contained such things) that all the satellites had observed the same cosmic event, and not some local source of noise or interference. The similarity of the three signals, measured by detectors roughly 100,000 miles apart, proved this. The "backgrounds" were the result of cosmic rays and radioactivity aboard the satellites; the left-hand parts of the figure were included to show that the gamma-ray burst signals were more than ten times stronger than the backgrounds. In addition, the backgrounds were nearly constant, while the gamma-ray burst fluctuated wildly, with several peaks in intensity indicated by arrows. The stretched (logarithmic) time axis is unusual, but was used for Vela data to make it possible to store in the satellites' very limited memory measurements of both rapidly varying brief events and more slowly varying longer events.

Some of the variations in the gamma-ray signal were just what scientists call "counting statistics": if you toss a fair coin 100 times you expect an *average* of 50 heads and 50 tails, but you are almost (61%) as likely to get 55 and 45, 92% as likely to get 48 and 52, and so forth. An exactly even split occurs in only about 8% of all series of 100 tosses, and in about 0.3% of all series of 10,000 tosses.

These statistical fluctuations are more important if there are fewer tosses. With a single toss you expect an average of one-half head and one-half tail, but can only obtain 0 and 1 or 1 and 0, which are not even close to one-half and one-half. With two tosses you have a 50% chance of getting 1 and 1, but a 50% chance of getting either both heads or both tails (in half of all two-child families the children are

Figure 1-1. The gamma-ray burst of August 22, 1970, as recorded by three Vela spacecraft. The count rates (the numbers of gamma rays recorded per second) are shown. The peculiar time axis was designed by the spacecraft builders so that both rapidly and slowly varying events could be studied even though the spacecraft could only store and transmit a very limited number of data. The vertical arrows show peaks observed by each of the three satellites, establishing that they are a genuine cosmic phenomenon and not some source of noise aboard one of the spacecraft. (*Astrophys. J. Lett.* V. 182, p. L87 [1973].)

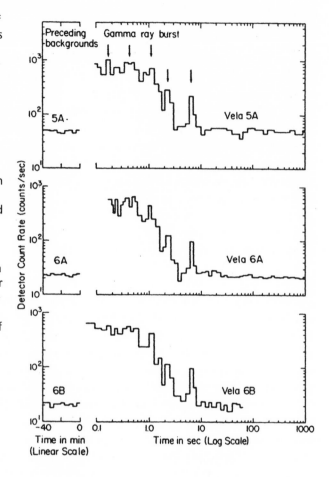

of the same sex). On the other hand, with 10,000 tosses it is more likely than not that the final count will be between 4950 and 5050 heads (and between 5050 and 4950 tails), differing from the expected count by less than 1%. Statistical fluctuations account for the irregularities in the background signals in Figure 1-1, obtained before the burst and after its end, when comparatively few gamma rays were counted.

Counting statistics are a source of experimental error introduced by the fact that real instruments are not perfect, and collect and count only a tiny fraction of the emitted gamma rays. Analogously, if you infer from 100 tosses that produced 48 heads and 52 tails that the coin was loaded, you would probably be wrong—you have no right to expect exactly 50 heads and 50 tails, even though that would be the *average* result for a fair coin. However, the peaks indicated by the arrows in Figure 1-1 were observed by all the instruments and are too large to be merely statistical fluctuations. They must be real variations in the gamma-ray intensity.

After about 8 seconds the signals shown in Figure 1-1 returned to approximately the background level they had before the burst began. As well as could be measured, the burst was over, although it was impossible to rule out the possibility that a low level of radiation continued much longer.

Not all gamma-ray bursts look like the one shown in Figure 1-1. For example, Figure 1-2, a page from a more recent catalogue of gamma-ray bursts, shows eight of them. They are together on this page only because they were observed consecutively (the trigger numbers, which are not consecutive, include a variety of phenomena other than gamma-ray bursts). The scales of the time axes are all different, sometimes by large factors. We see that gamma-ray bursts are a very mixed bag. Some last more than a minute, others less than a tenth of a second. Some consist of a single pulse, smooth except for the effects of counting statistics (the first in the right column, or the second in the left column), or two pulses (the third in the left column). Others consist of a large number of clearly separated subpulses (the last in the right column contains at least eleven distinct subpulses). Sometimes the intensity drops to background levels between subpulses (the last in the right column), meaning that no radiation from the burst is being detected at all, while in others narrow spikes arise out of an elevated level of steadier emission (the second in the right column shows both these phenomena). This diversity has been summarized in the maxim "When you've seen one gamma-ray burst, you've seen one gamma-ray burst." It was clear from the first that any explanation of gamma-ray bursts must be able to accommodate an enormous variety of behavior. This variety, rather than any specific pattern, was the clue to be found in these data.

Gamma rays travel in straight lines at the speed of light, and in the vacuum of space they are neither absorbed nor scattered. Because they cannot be bent by lenses or reflected by mirrors it is impossible to focus them, or to form a gamma-ray image or picture like those made by an ordinary camera or telescope. The instruments that detect gamma rays work by absorbing their energy, and are generally incapable of determining from what direction they arrive. They can, however, measure their time of arrival very accurately. As a result, if a burst of gamma rays from a single source is observed by more than

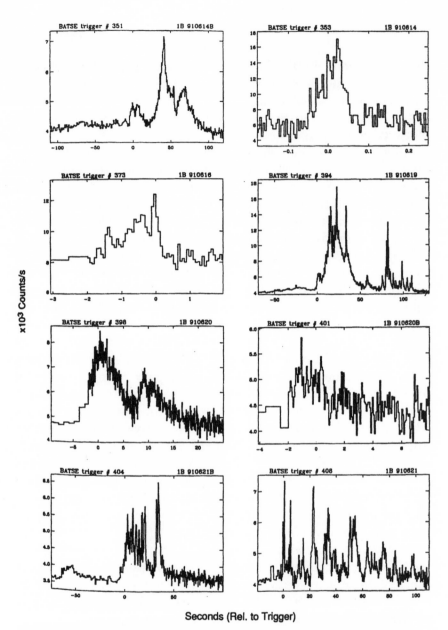

Seconds (Rel. to Trigger)

Figure 1-2. A few exhibits from the zoo of gamma-ray bursts. Note the differing time axes; some bursts last more than a minute, others less than a tenth of
a second. Some consist of many separate subpulses; others contain only a single pulse. Some have sharp spikes on top of more smoothly varying pulses. In
every case the data show statistical fluctuations as random and irregular variations from one time bin to the next; these fluctuations are larger when the
count rate is low and when the time resolution is great (each bin in the plot
represents only a very short interval, so the total number of gamma rays detected is small, such as for triggers #353 and #401). The eye is good at
smoothing out the statistical fluctuations. Real variations in intensity are consistent from one time bin to the next; most of the spikes in triggers #394 and
#408 are probably real. Trigger #398 probably has only two broad peaks; all
the rest of its "ups and downs" are statistical fluctuations. (*Astrophys. J. Suppl.* V.
92, p. 251 [1994].)

one detector, it is possible indirectly to determine, or at least to constrain, the location of their source.

The principle is shown in Figure 1-3, if there are two detectors, D1 and D2. A burst at S1 will be observed at the same time by both detectors. So will a burst occurring anywhere on the dashed line. On the other hand, D1 will observe a burst at S2 later than D2, by an amount equal to the distance between these two detectors divided by the speed of light. The same will be true for a burst occurring anywhere on the dotted line. Bursts from intermediate directions produce a lesser time delay. Thus, with two detectors it is possible to determine the direction on the page from which the gamma rays arrive simply by comparing the signal arrival time at the two detectors, even though the detectors individually are completely incapable of determining the arrival direction of the gamma rays. This method does not work with steady sources of radiation, but is well adapted to brief sources of gamma rays, such as nuclear bombs and cosmic gamma-ray bursts.

In three dimensions, two detectors can determine that a burst came from somewhere on a thin curved sheet in space. If the source is far from the detectors, this sheet is the surface of a cone. Three detectors can generally localize a burst to a line in space, and four to a single point. The four detectors form a very large but very incomplete synthetic lens—large because the detectors are spread over such

Figure 1-3. Determining the direction of a gamma-ray burst using a network of satellite detectors (the Vela system, or the later interplanetary networks). A pulse of radiation emitted by a source S1 (or anywhere on the dashed line through S1) arrives at the detectors D1 and D2 at the same time. A pulse emitted by a source S2 (or anywhere on the dotted line through S2) will be received by detector D2 before it is received by detector D1 with a time difference equal to the time it takes the signal to travel the extra distance between D2 and D1. Sources in other directions will be detected with other time differences. The measured values of the time differences can be used to determine the directions to the sources, but not their actual distances (unless they are very close).

a large region of space, and incomplete because they sense only a tiny fraction of the gamma rays passing through that region. The Vela system generally had four satellites operational for this reason. It is also one of the reasons why the satellites were in such high orbits: the further apart the detectors, the more accurate the localization, just as a large telescope may produce sharper images than a small one.

The focusing ring on a camera has distances marked. On my camera (an Olympus OM-1) the markings are 0.45, 0.5, 0.7, 1, 1.5, 2, 3, 5, and 10 meters, and ∞ (infinity). The entire range of distances above 10 meters, out to 100 meters, 1 kilometer, the Moon at 400,000 kilometers, and on to the most distant stars, is compressed into a tiny sliver of the focusing ring. They are all "at infinity," as far as the camera is concerned, though certainly not if one tries to pace them out. The reason is that what matters for the optics is not the distance but the reciprocal of the distance. The reciprocals of 100 meters, 1 kilometer, the distance to the Moon, and the distance to the farthest stars are all essentially zero, to the accuracy of the camera's optics.

Determining the distances to gamma-ray bursts from their arrival times at several detectors works in roughly the same way. The designers of the Vela system chose the size of its orbits to be able to find the location of a nuclear explosion out to several times the distance of the Moon. This roughly corresponds to the distance at which a small nuclear explosion would produce enough X rays and gamma rays to be detected at all. Further out, the Vela system was only able to set a lower bound on the distance to a burst of gamma rays. Like a camera, it could not distinguish an object at the distance of Mars from one in the most remote galaxy.

Once the Vela team realized they were observing remote astronomical events they did not need to determine distances from the timing data. It was sufficient to assume they were very far away ("at infinity," as a photographer would describe it). With this assumption, only three observations are required to determine the direction to the source of gamma rays. If they insisted on using four observations to determine the distance too, all they found was a lower limit of a million miles or so. The bursts could not be coming from Earth, and directional information also ruled out the Sun and planets. Other than those two obvious candidates, almost any location in the universe was permitted by the data, from within the solar system to past the most remote galaxies.

Because the timing information obtained by the Vela satellites was not extremely accurate (the original publication quotes an accuracy of ±0.05 second) the directions of the gamma-ray bursts also had a substantial uncertainty, typically a few degrees of arc. This is several times the angular size of the Moon and the Sun, and is very crude by the standards of astronomy, in which the positions of stars are measured to fractions of an arc-second (an arc-second is $\frac{1}{3600}°$). It precluded identifying gamma-ray bursts with any other astronomical object. A swath of sky several degrees across is much too big to search in detail; it would be like looking for a single blade of crabgrass in an acre of lawn, if you had no inkling of what crabgrass looks like or even if it is visible at all.

Although individual bursts could not be associated with any individual astronomical object, data from all the observed bursts could be combined to make statistical inferences about their origin. Using the measured directions to nine bursts (and more limited directional information about an additional ten bursts, each of which was observed by only two satellites) the Vela team concluded that gamma-ray bursts were equally likely to be observed from any point on the sky; this is called an isotropic distribution. This conclusion is now supported by observations of thousands of bursts; no statistically significant deviation from isotropy has ever been found.

There could be no doubt that gamma-ray bursts were real. They were also completely unexpected, and completely mysterious. Solving the mystery required the development of a series of new instruments that could make new kinds of measurements. One task was to determine where in the universe gamma-ray bursts come from. Another was more theoretical, to understand how they work. Not every observational clue, or new idea, led in the right direction; some were wrong, or simply led to dead ends. No one anticipated how long it would take.

2

Detectors

In 1608 the Netherlandish eyeglass maker Hans Lippershey put two lenses together and invented the telescope. He intended it as an instrument of war, for spying out the enemy. The Estates of Holland, in the midst of the Eighty Years' War of independence from Spain, awarded him a substantial sum.

New ideas moved fast in those days, and were put in practice as quickly as they are today, even though mail traveled by horse. Within a year Galileo Galilei, professor of mathematics at the University of Padua, in Italy, and Thomas Harriot, an English mathematician, heard of Lippershey's invention, built their own telescopes, and turned them to the sky. Galileo discovered the phases of Venus (like those of the Moon) and mountains and craters on the Moon, resolved the Milky Way into individual stars, and began the systematic study of sunspots. Both men independently discovered the moons of Jupiter. Like gamma-ray bursts, these discoveries were an unanticipated spin-off of a military technology. They proved the Copernican picture of the solar system and rescued astronomy from the realm of theology and philosophy, placing it firmly in the land of science.

Astronomy, as a science in the modern sense, was born with the telescope. Science advances when technology makes new experiments possible, and only as far as technology makes possible. Science is a creature of technology, in our own day just as in Galileo's.

The scientist who wishes to study gamma rays, whether to do nuclear physics in the laboratory or gamma-ray astronomy in space, must first detect them. To do this he or she must study the technology of gamma-ray detectors. The possible detectors are determined by the physical properties of gamma rays, the kind of data the scientist wishes to obtain, and the environment in which he must work.

The most important property of gamma rays is that they penetrate matter well. Just how well depends on the energy of the individual gamma ray and the nature of the matter. A second, and almost as important, property is that their interaction with matter is all or nothing: either they pass through with no interaction, or they leave most or all of their energy, giving it to an electron or sharing it between an electron and a positron (the electron's anti-particle, identical to an electron except that its charge is positive rather than negative). In contrast, alpha rays (the nuclei of helium atoms, produced when uranium and other very heavy elements decay) and beta rays (energetic electrons) leave energy in a continuous trail along their paths. It is the difference between cardiac arrest (gamma-ray interaction) and gradual exhaustion (alpha rays and beta rays).

Gamma rays are electromagnetic radiation, consisting of oscillating electric and magnetic fields, just like visible light, radio waves, and X rays, but are much more energetic. The energies of gamma rays are measured in units called electron Volts (eV), with 1000 eV abbreviated as KeV and 1,000,000 eV abbreviated as MeV. An electron Volt is the energy an electron (or any other singly charged particle, such as a positron or a proton) would acquire in traveling without friction between two surfaces whose voltage difference is 1 Volt.* Electrons are the elementary particle that has almost all the negative charge in the universe, carries current through wires, and determines the chemical properties of atoms and molecules.

A photon (the indivisible fundamental unit) of visible light has between 1.8 eV and 3 eV of energy, depending on its color, while a radio photon has thousands or millions times less. X rays are typically defined by physicists as photons with between about 100 eV and about 100 KeV. Gamma-ray astronomers are usually concerned with photons with energies of 30 KeV or more, a definition based on the type of detector used to observe them. The dividing line between

---

*If an electron were to travel without energy loss between two wires plugged into an ordinary wall outlet, it would acquire an energy of 120 eV, reflecting the familiar electrical supply at 120 Volts (in the United States). Because the voltage is alternating (AC), varying at a frequency of sixty times per second, this is only an average value. When electrons actually flow through wires they suffer frequent collisions and never reach this maximum energy, but in the vacuum of a television tube, or interstellar space, they do achieve the maximum energy permitted by the accelerating voltage— about 20 KeV for the picture tube, and enormously more in some astronomical objects.

gamma rays and X rays is arbitrary and the definitions overlap. X-ray and gamma-ray astronomers are forever poaching on each other's domains.

Both nuclear physics and gamma-ray-burst astronomy are chiefly concerned with gamma rays of energies between about 30 KeV and 3 MeV. An energy of 300 KeV is halfway between these two limits on a logarithmic (ratio) scale, and is in the middle of the range containing most of the radiation of most bursts. This is the most important energy for a gamma-ray-burst detector. Gamma rays of this energy will, on average, penetrate hundreds of feet of air, about 3 inches of water, or about two-thirds of an inch of a crystal called sodium iodide, before being absorbed. For a detector to be efficient it must detect most of the gamma rays that fall on it. Gamma-ray bursts, as powerful as they are, occur so far from us that our detectors must make efficient use of the few gamma rays that do reach us.

The ability of a material to absorb gamma rays depends on two factors. The first is simply its density—the more mass ("stuff") there is, the more gamma rays will be absorbed. Air, and other low-density gases, are therefore very inefficient absorbers of gamma rays and cannot make good detectors. This rules out devices like the traditional Geiger counter, which senses energy deposited in a low-density gas. These work well with beta rays, which are not nearly as penetrating and leave a continuous trail, but are terribly inefficient detectors of gamma rays. Even if a beta ray leaves only a tiny fraction of its energy behind, a Geiger counter will detect that energy, but it will completely miss almost every gamma ray that passes through it.

The second property of a material affecting its interaction with gamma rays is the atomic number—the number of electrons per atom—of the atoms it contains. Elements of higher atomic number are better absorbers. This is partly because each atom contains more electrons, but also because each electron has a greater chance of absorbing a gamma ray if it is close to a nucleus of high atomic number.

The classic detector of gamma rays is called a crystal scintillator. It was developed in 1947, as the nuclear age led to a great flowering of nuclear physics. At its heart is a large crystal of sodium iodide, doped with a small quantity of the heavy metal thallium, and denoted NaI(Tl). The chemical symbols Na and I stand for sodium and iodine, and Tl for thallium. The Tl is in parentheses because it is only added in tiny quantities and does not change the basic chemistry of the crystal.

Sodium iodide is a close chemical cousin of table salt, sodium chloride. Each is a transparent colorless crystal. The chief reason for substituting iodine for chlorine is that the atomic number of iodine (53) is much greater than that of chlorine (17). Iodine is a much better absorber of gamma rays. Sometimes, as in the Vela satellites, another close cousin, cesium iodide [CsI(Tl)], is used in place of sodium iodide, for the same reason; the atomic number of cesium (55) is much greater than that of sodium (11).

When a 300-KeV gamma ray is absorbed by a sodium iodide crystal it tears one of the electrons out of the sodium or iodine atoms. Sometimes the energy of the gamma ray is completely given to the electron (a process called photoelectric absorption), but sometimes a new gamma ray of lesser energy is emitted simultaneously (a process called Compton scattering), with the difference in energy, usually a substantial chunk of the initial gamma ray's energy, given to the electron. Electron-positron pairs can be created only if the initial gamma ray has an energy of at least 1.022 MeV.

The liberated electron (which in other circumstances would be called a beta ray) carries quite a bit of energy by atomic standards and rips through the crystal like a bullet through flesh. Because it is negatively charged, it repels electrons in its path, pushing some of them out of their atoms, and leaving behind a continuous trail of freed electrons and damaged atoms (called positive ions because they have lost an electron). This process is termed secondary ionization. The interaction between the energetic electron and the crystal is strong, and it does not go far before losing its energy.

In a few millionths of a second the free electrons find their way back to the positive ions from which they were stripped, and the wound heals. On their way, some of these electrons bump into thallium atoms, giving them a little of their energy, and raising them to an excited (more energetic) state. The excited thallium atoms give up this energy by emitting visible light. That is why they were added.

This entire process is called scintillation, because it makes a flash of visible light from the gamma ray's energy. NaI(Tl) and its relatives are transparent, so this light travels freely throughout the crystal, escaping from its surfaces. It may be detected by a photomultiplier tube (a vacuum tube that makes an electric signal from light) butted up against the crystal, or a wide fiber-optic light pipe may collect the light from a large crystal and direct it to a photomultiplier tube.

NaI(Tl) scintillators have a number of desirable properties that account for their use in most gamma-ray-burst detectors. They are reasonably cheap, rugged, absorb gamma rays efficiently, produce easily detectable light pulses, are sensitive to gamma rays arriving from any direction, are available in large sizes (discs 20 inches in diameter, the size of a laundry basket, have been used), do not deteriorate with time, require no maintenance, and work well at any temperature likely to be encountered. These properties are useful in the laboratory and essential in space.

NaI(Tl) scintillators have two chief disadvantages. One is that the direction of the electron that carries the gamma ray's energy is only roughly related to the direction from which the gamma ray came. In addition, unless the gamma ray was very energetic (many times the 300 KeV the gamma-ray-burst astronomer is detecting) it is impossible to measure the direction of this electron because it runs out of energy near where it was produced, deep within the scintillator. By the time the emitted light is detected by the photomultiplier tube all directional information has been lost. This sensitivity to gamma rays arriving from any direction is essential to detecting events, such as gamma-ray bursts or clandestine nuclear explosions, which are unpredicted and whose direction is not known in advance, but it is difficult, afterward, to determine which direction the gamma rays came from.

The laboratory nuclear physicist does not mind this; he generally knows from the design of his experiment in exactly what direction his gamma rays are traveling. It is, however, a serious deficiency for the astronomer. This property of scintillators is exactly opposite to that of optical telescopes, which view only a tiny fraction of the sky at any one time, but measure the directions of the objects they see extremely accurately. Gamma-ray telescopes are not really telescopes because they do not form an image or picture from the radiation they detect. That is why they are generally called detectors, a term borrowed from the nuclear physics laboratory, rather than telescopes, even when used by astronomers. Bridging the gap between these two very different kinds of instrument has been a central theme of gamma-ray-burst astronomy, and the central task of the observer of gamma-ray bursts.

The second disadvantage is that NaI(Tl) scintillators give only a rough idea of the energy of the gamma rays they detect. A more energetic gamma ray usually gives a more energetic electron, more secondary ionization, and more light. Unfortunately, there are several difficulties in extracting the gamma-ray energy from this information. One is that some of the initial gamma ray's energy may be carried off by a second gamma ray. More seriously, the inexorable mathematics of counting statistics (Chapter 1) applies to the production of visible light. As a result, the energy of a 300-KeV gamma ray is only measured to an uncertainty of about ± 20 KeV, or $\pm 7\%$ (in more technical language, the standard deviation of the energy measurement is 20 KeV). This problem gets proportionately more severe at lower energies, roughly as the square root of the gamma-ray energy, so that for a 30-KeV gamma ray the uncertainty is about ±6 KeV, or ±20%. Even worse, the light pulse measured by the photomultiplier tube depends to some extent on where in the crystal scintillator the gamma ray was absorbed, for the same reason that a lamp may not illuminate the corners of a room as well as its center. A more energetic gamma ray may therefore masquerade as one of somewhat lesser energy, or vice versa. This effect is difficult to calibrate and usually impossible to remove from the data.

Nuclear physicists also had to deal with these difficulties because they are intrinsic to the materials of crystal scintillators. The laboratory physicist can often use compact crystals, which mitigates the last problem. For example, one standard size is a cylinder 3 inches in diameter and 3 inches long. In contrast, the astronomer must use very wide (but comparatively thin) laundry-basket-sized scintillators to maximize the detecting area and the number of gamma rays collected from a weak source.

The nuclear physicist has an even better solution. He uses an entirely different material, the crystalline semiconductor germanium (Ge), a close cousin of the silicon (Si) used in most integrated circuits. In a Ge detector the ionization is detected electronically, rather than by sensing visible light. This improves the accuracy of measurement of the gamma-ray energy at least tenfold. As a result, Ge detectors have largely replaced NaI(Tl) in the nuclear physics laboratory.

Unfortunately for the gamma-ray astronomer, Ge detectors have two deficiencies that make them very hard to use in space. One is that they must be cooled to −150°C, or colder. This is easy in the laboratory, where cooling is generally provided by liquid nitrogen, boiling at −196°C, and costing about 10 cents per liter. But liquid nitrogen gradually boils away. Resupplying a satellite usually costs as much as replacing it entirely, and sometimes much more. Beyond low Earth orbit, resupply has never been attempted. In order to operate low-temperature astronomical instruments in space the entire space-craft must be designed around either a large Dewar flask (thermos bottle) carrying the irreplaceable coolant, a cumbersome refrigerator, or a device known as a passive radiative cooler. This consists of thermal shields that shadow the instrument from both sunlight and the warmth of Earth. It cools by radiating its own heat, in the form of infrared radiation, to the cold darkness of space, just as you are chilled in a room with cold walls, even if the air is warm. Cooled detectors have been used in satellites studying infrared and millimeter wave radiation, such as the cosmic microwave background, for which there is no alternative, but only a few such low-temperature gamma-ray detectors have been launched.

The second difficulty with Ge detectors is that they are not available in the laundry-basket sizes necessary to detect astronomical sources of gamma rays, which are generally quite faint because of their distance. A laboratory nuclear physicist can usually circumvent this problem by placing his detector close to his source of radiation, using a strong source, or accumulating data from a steady source for a long time. The gamma-ray-burst astronomer is stuck with the faint and brief sources nature provides. The few Ge gamma-ray-burst detectors that have been launched have been disappointments because they have not detected enough gamma rays, even from the brightest bursts, to produce an accurate spectrum.

New electronic materials are being developed to solve these problems. In particular, a semiconductor known as cadmium zinc telluride (CZT), divided into a large number of tiny sensing elements by the techniques used to make integrated circuits, will make laundry-basket-sized detectors that also measure the energy of the arriving gamma rays nearly as well as a Ge detector. Even better, by placing the detector behind an absorbing mask and measuring the position of the mask's shadow on the array of sensing elements it will be possible to measure simultaneously the direction of arrival of the gamma rays. This is like determining the time of day from the length of your shadow on the ground. Unfortunately, CZT detectors have been decades in the making. NASA does not support the headlong pace of technical progress routine in the electronics and computer industries. The eventual triumph of gamma-ray-burst astronomy was founded on ingenuity in using existing technologies, not on better detector materials.

The engineers who designed the Vela satellites used CsI(Tl) detectors and faced, in the 1960s, the same problems gamma-ray-burst astronomers had to deal with in subsequent decades. The Vela designers were not terribly concerned that the energies of the gamma rays detected were not accurately measured because it would have been sufficient for their purposes to establish the existence of a sudden burst of gamma rays. It would not have been necessary to know the energies of the gamma rays accurately to establish their source because nothing else resembles a nuclear bomb. The Vela instruments did produce rough energy measurements, but these did not need to be of high quality to fulfill the mission of treaty monitoring. The Vela satellites also measured the energies of the gamma rays from gamma-ray bursts, but these data were not very informative.

The Vela engineers faced, and solved, the problem of determining the direction of a sudden burst of gamma rays using instruments that, individually, were completely indifferent to the direction the gamma rays came from. As we saw in Chapter 1, they did this by measuring the burst arrival times at several, widely separated, detectors. Gamma-ray-burst astronomers developed this method further.

Not long after the discovery of gamma-ray bursts a tacit consensus arose that the solution to their mystery lay in identifying bursts with astronomical objects observed at other wavelengths, especially visible light. Such identifications would require accurate positions of at least a few bursts. This consensus was spontaneous, rather than the result of vigorous public debate, because this approach had successfully solved the problems of radio and X-ray astronomy. It was also a straightforward path, using the principles and methods of the Vela developers, given the available detector technology. Simply distribute a minimum of three gamma-ray-burst detectors as widely as possible throughout the solar system and record the arrival times of the gamma-ray bursts. In principle, these data could be collected, analyzed, and distributed as fast as radio waves, traveling at the speed of light, could bring data from the widely dispersed satellites. This might take a quarter of an hour. If everything worked smoothly, an optical telescope could be pointed toward the burst soon after it happened. In practice, especially in the earlier days, collecting and analyzing the data took much longer.

These systems are generally called interplanetary networks (IPN). There have been several such networks since the late 1970s, but various mishaps and spacecraft failures have meant that no network existed during most of that period. The accuracy of the gamma-ray-burst positions that can be obtained is proportional to the separation of the satellites (the angular uncertainty is proportional to its reciprocal). By expanding beyond the near-Earth distribution of the Vela satellites a thousandfold improvement in accuracy might be obtained, reducing the area on the sky to be searched a millionfold. The accuracy also depends on how sharp a peak can be identified in the time history of the gamma-ray intensity, and hence on the shape of this curve, the intrinsic time resolution of the detectors and their data recording, and the accuracy of their clocks. In practice, the more accurate positions typically had uncertainties of several arc-minutes (an arc-minute is $\frac{1}{60}°$), roughly a hundred times better than the Vela system had provided.

Gamma-ray-burst detectors in the IPN were piggybacked on satellites launched for other purposes. They had to be cheap, small, and light, and could only detect a few of the brightest bursts. Searches of the sky around burst positions measured by the IPN were fruitless, with the sole exception of the burst on March 5, 1979 (Chapter 6). Any steady counterpart to gamma-ray bursts must be very faint. This only deepened the mystery.

3

Where Are They?

The hardest problem in astronomy is to find the distance to an astronomical object. It is not possible to pace out the distance or extend a tape measure. Radar has been used to determine distances in the solar system by measuring the time required for a radar pulse, traveling at the speed of light, to bounce off an object and return to us. This method does not work outside the solar system because the time required for the pulse to return would be too long (8 years for the nearest star), and because the returned signal would be undetectably faint.

If the intrinsic properties of an object are known, then its distance may be estimated from its appearance. For example, we judge the distance to a person from how big he appears, and the distance to a hill from the apparent size of trees or houses on it, whose actual size we know. This method is so familiar that we use it without thinking about it.

If we know how luminous a source of light is (its total light output), its apparent brightness may tell us how far away it is. Although commonsensical, this method does not work well in everyday life. Our eyes do not measure brightness accurately, and the brightnesses of many artificial sources of light (such as a flashlight or headlight) depend very much on how they are pointed. A distant powerful flashlight pointed toward us may appear as bright as a much closer weak flashlight, and a flashlight pointed away from us is almost invisible. These everyday sources of light are not "standard candles," the astronomers' term for an object of known luminosity radiating equally in all directions, taken from the nineteenth-century physicists' laboratory standard of luminosity (an actual candle, made in a standard-ized and carefully controlled way).

Astronomers can determine the distances of ordinary stars by mea-suring their apparent brightnesses because instruments do this accu-rately and because stars radiate nearly equally in all directions. This method could not be used until the properties of stars were accurately determined, well into the twentieth century. It cannot be applied to newly discovered or poorly understood phenomena, such as gamma-ray bursts, whose luminosity is unknown.

The basis of all astronomical distance measurements outside the solar system is parallax: the direction to an object depends on the place from which it is viewed. Even your two eyes view from slightly different locations, so that nearby objects appear to jump when you close or cover one eye, looking with the other eye, and then switch eyes. When we have both eyes open, our brains use parallax to con-struct a three-dimensional image of the world because the parallax is inversely proportional to the distance of the object. Binoculars have their lenses set wide apart, effectively increasing the separation of our eyes, to increase the parallax of distant objects.

Astronomers view stars twice, 6 months apart, as if forming a giant pair of binoculars with lenses separated by the diameter of Earth's orbit. Even with this enormous separation the parallaxes of stars are tiny—less than 1 arc-second—and difficult to measure.

Attempts to measure the parallax of objects outside the solar sys-tem go back at least to 1572, when the Danish astronomer Tycho Brahe tried, and failed, to measure the parallax of the bright super-nova (exploding star) of that year. Now we know that Tycho's super-nova was so distant its parallax could not be measured even with today's best instruments, but his efforts were important, both for the invention of the method and because even his negative result, ob-tained from naked-eye observations, demonstrated that the super-nova, like the stars, lay outside the solar system. Friedrich Bessel, in 1838, was the first to measure the parallax of a star. His work was the first actual measurement of the distance to a star other than the Sun, whose distance had been known since ancient times (also measured by parallax, using observations from two points on Earth).

After using parallax to determine the distances to the closest stars, astronomers were able to calibrate the properties of various types of stars, which then serve as standard candles. If one such calibrated star is found in a cluster or galaxy, the distance to the entire group of stars is known and can be used to calibrate the luminosity of another, rarer but more luminous, type of object. The more luminous a standard candle is, the further it can be observed and used to measure distances.

Even supernovas, the brightest objects in the universe other than gamma-ray bursts, can be used as standard candles. The supernova of 1885 in the Andromeda Galaxy was the first clue that this "nebula" is actually a galaxy roughly as luminous, large, and massive as our Milky Way. Today, supernovas, observed in the far reaches of the universe, have become a fundamental tool of cosmology, used as standard candles to study the acceleration of its expansion and to measure Einstein's cosmological constant.

None of this helped the gamma-ray-burst astronomer. Parallax requires accurate measurements of the position on the sky of a source of radiation, precisely the kind of data unobtainable from gamma-ray detectors. Even had this not been an obstacle, it would have been impossible to measure the parallax of an event that lasted, at most, a few minutes by a method that depends on observations obtained months apart. Aside from these insuperable difficulties, we now know that gamma-ray-burst parallaxes are a million times too small to measure.

Direct geometrical information about the distances to gamma-ray bursts consisted only of the lower bounds of about a million miles produced by the Vela satellites. Interplanetary networks of gamma-ray-burst detectors eventually increased these lower bounds to ten, or even a hundred, times the distance from Earth to the Sun. The inner solar system was excluded, but the outer solar system was not.

Greater distances were also possible, and more plausible; astronomers regard the solar system as familiar turf, just as a street gang regards its immediate neighborhood, and are loath (with somewhat more justification than the gang) to admit that it may contain interlopers, such as the mysterious sources of gamma-ray bursts. The Sun resides in a vast pancake of stars, dust, and gas known as the galactic disc (so called because it is wide, thin, and rotates, roughly like the discs used for storing information in computers, music, and the newer video devices). Most of the stars in our galaxy are found in its disc, which is about 100,000 light-years in diameter and 1000 light-years thick, although it has no sharp boundary, but rather extends, with gradually decreasing density, in all directions. The Sun is about 30,000 light-years from the center, but lies almost exactly in the midplane of the pancake.

When a physicist or astronomer does not know the value of a quantity he tries to estimate its scale by comparing it to known quantities. For example, if someone is in the same room with you, but you don't know where in the room, you might estimate that person's distance as 10 feet because most rooms are between 10 and 20 feet long and wide, and two people randomly placed are typically about 10 feet apart. This estimate would be quite wrong if, for example, your friend were at the opposite end of St. Peter's in Rome, several hundred feet away, or if you were locked in intimate embrace. Without additional information 10 feet is a fair estimate, the best estimate you can make, and, most important, is likely to be close enough to the actual distance to be useful. Scientists call this making an order of magnitude estimate or setting the scale of a problem; occasionally it is wrong, but more often it is a useful guide to thinking. Most of the time we are not in St. Peter's, and even ardent lovers are not always in each others' arms.

Astronomers faced with the discovery of gamma-ray bursts first asked from where in the universe they might be coming. The various regions of space may be labeled on the basis of an order of magnitude estimate of their distances. For example, the inner solar system is that volume within a few times the Earth–Sun distance, while the outer solar system may be defined as the region at least ten times as far from us as the Sun, but no more than ten thousand times as far. Beyond that is interstellar space. Out to a few hundred light-years is the solar neighborhood, and from that to 100,000 light-years is the galaxy. After that is intergalactic space, with the first several million light-years belonging to the local group of galaxies, and then the more distant universe. Finally, distances of billions of light-years are called cosmological, because they include the entire universe, and light from these most remote regions takes so long to reach us that it was emitted when the universe was significantly younger than it is now and had different properties.

Of course, all these more distant regions contain large numbers of planets, stars, and galaxies, so that all the features of our solar system, galaxy, and other nearby regions are reproduced, both in familiar and different forms, an enormous number of times, just as foreign countries contain families, towns, and cities like ours. In addition, a great volume will also contain a few rare objects that cannot be found nearby. You must travel to a good art museum to find an original Rembrandt, and the nearest quasar is many hundreds of millions of light-years away.

In general, the nearer a region the more closely it can be studied and the better its contents are known and understood. The more distant an object the more energetic and luminous it must be to be observed and the harder it is to explain its energy release. Astronomers are therefore torn between trying to ascribe a new and ill-understood phenomenon to nearby familiar objects or to distant and exotic ones. It might seem easier to assign it to the dark mysteries of remote space, but in fact astronomers have generally been prejudiced in favor of less distant origins. For example, although quasars were conclusively shown to be at extremely great (cosmological) distances by the mid-1960s, this conclusion was accepted only with reluctance and a few die-hards insisted for decades, against all evidence, that they were much closer.

The observed brightness of an object varies as the inverse square of its distance (at twice the distance it is one-fourth as bright, etc.). This is known as the inverse square law, and follows from the geometrical fact that the surface area of a sphere, over which its radiation is spread, is proportional to the square of its radius. Therefore, assuming a great distance for an object of known brightness requires that it be enormously luminous. This must be balanced against the fact that the further out in space you look, the more possible types of objects and phenomena, including very rare ones, there will be.

Assuming a distance scale for a newly discovered phenomenon also has unavoidable implications for the directions in which it is observed. For example, if a phenomenon is assumed to occur near the Sun, then it must be observed in that direction. Usually, the implications are not quite so obvious, but they may nevertheless be a powerful constraint. The new phenomena are likely associated with, and therefore lie in the same directions as, previously known objects at roughly the same distance. This may sound arbitrary, but it is true if the new and old objects are related, either directly or by a common origin or ancestor, and it has often been borne out by experience. Ockham's razor, the principle that the world is not unnecessarily complicated,* points us toward the assumption that newly discovered phenomena are a consequence of previously known phenomena.

For example, the more distant stars of the Milky Way form a broad band across the sky, lying in the plane of the galactic disc. Many known classes of astronomical objects also have this distribution, in-

---

*Named for William of Ockham (1285–1349?). Why is this product of medieval philosophy useful in modern science? Probably because if you have license to construct complicated explanations, you can appear to explain almost anything by introducing complications whenever your simple ideas fail, even if you really don't understand it. A simple explanation will work only if it is right. Of course, Ockham himself was thinking of philosophy, and the scientist who applies this principle is taking it out of its original context.

cluding pulsars, supernova remnants, steady gamma-ray sources, X-ray sources, and clouds of neutral hydrogen and molecular gases. In fact, these objects are related because most of them are comparatively (on astronomical time-scales) short-lived products of the evolution of massive stars, which are themselves produced by dense clouds of neutral hydrogen and molecular gases. Naturally, all of these are found in the same places in the galaxy, because their lives are too short and speeds too slow to travel far from their birthplaces.

It was evident from the first that gamma-ray bursts are not preferentially found in the band of the Milky Way. This did not prove that they were unrelated to massive stars, only that they were not related to the massive stars of our galaxy or their neighbors in the galactic plane, or that they had somehow traveled far from these objects.

Other distance estimates lead to other predictions for the distribution of bursts on the sky. Distances within the solar system would generally imply a concentration in the plane of the planetary orbits, near which most of the components (except for comets) of the solar system are found. Similarly, our galaxy contains a halo of old stars, some moving by themselves and some bound into globular clusters each containing hundreds of thousands of stars. This halo is centered on the center of the galaxy. The Sun's off-center location means that from Earth any population of objects associated with the galactic halo will be roughly concentrated toward the galactic center. This is analogous to the fact that when you are outside a large city its light produces a glow in the nighttime sky in its direction. Both the solar system and the galactic halo, along with the galactic disc, are thus excluded as the origin of gamma-ray bursts.

The simplest possible distribution is an isotropic one, in which a gamma-ray burst is equally likely to come from any direction. This is consistent with many possible source populations. One, rather far-fetched (because no one ever invented a plausible explanation of how they can make gamma-ray bursts), is the cloud of comets that fill the outer solar system. Another, more plausible, is the solar neighborhood, interstellar space at distances less than the thickness of the galactic disc. If you are only observing such comparatively close objects, the galaxy's pancake shape is not evident, just as a small ant eating its way through a real pancake, unable to feel beyond the ends of its tiny antennæ, cannot immediately tell the pancake's shape; as far as it can feel, it is batter in all directions. This is why the naked-eye stars, close enough to us to be seen without a telescope, are not restricted to the band of the Milky Way but are roughly isotropically scattered on the sky.

Yet another possibility is the distant reaches of the universe, so far from our galaxy and its neighbors that the matter appears to be uniformly distributed in all directions. This is a logical explanation of the observed isotropic distribution of arrival directions of gamma-ray bursts. It was perhaps the most exciting possibility, for the implied great distances would mean that gamma-ray bursts are extraordinarily energetic. Although the energy requirements would be forbidding, there would be a silver lining: if gamma-ray bursts are observed from very great distances, they must be very rare. If all the detectable gamma-ray bursts, about a thousand each year, are produced in our galaxy, then it must make them frequently, but if we observe bursts from the entire universe, containing perhaps a billion such galaxies, then each galaxy need only produce a burst once per million years. Rare and extraordinary processes may then be considered as explanations. It is easier to explain an extraordinary event if you don't have to do it very often.

Astronomers faced with the discovery of gamma-ray bursts considered all possible distances, from the solar system to cosmological. Perhaps because of the pervasive prejudice that less energetic explanations were more likely than more energetic ones, cosmological distances were among the last to be suggested. They were first mentioned in print in a review, presented at the end of 1974, of the initial flurry of interest that followed the discovery of gamma-ray bursts. The author, the Columbia University astrophysicist Malvin Ruderman, made a list of various possible sources for gamma-ray bursts, each with its determined advocates. He did not ascribe the suggestion of cosmological distances to any individual; it had apparently been considered, and discarded as implausible, by many. In fact, Ruderman presented a theoretical argument, now believed to be invalid because of an unduly low estimate of the characteristic energy of the gamma rays emitted by a gamma-ray burst, against the enormous energy output such great distances would imply.

The first explicit advocacy of cosmological distances appears to have been in two papers by the Soviet astrophysicist Vladimir Usov and his colleagues, published in 1975. Their suggestion was remarkably prescient, but although it was not completely forgotten, it received undeservedly little attention. This may have been because no plausible mechanism was apparent for explaining the enormous energy output implied by these distances. In addition, there were several apparently persuasive arguments that gamma-ray bursts were much closer.

Usov's suggestion may have been neglected for less valid reasons. Although scientists "officially" assign credit and priority on the basis of publication in professional journals, attention is more likely to be paid to face-to-face conversations, invited talks, and a gossipy grapevine. It Is a natural human failing to give credit to oneself and one's friends and associates while ignoring those outside these circles. Vladimir Usov is rather soft-spoken, and not a forceful advocate for his ideas, at least in English. He never became a member of the informal club of "leading" astrophysicists who organize meetings and invite each other to lecture at them. The Soviet Union restricted foreign travel; later, travel funds were scarce. He was not on the itinerary of most astronomers working on gamma-ray bursts. Despite this, Usov probably contributed more than any other theorist to unraveling the gamma-ray-burst puzzle, both establishing their distance scale and, later, explaining how they work.

4

What Are They?

The Double Helix, James Watson's account of how he and Francis Crick discovered the structure of DNA, contains a picture of Watson and Crick standing by their model of DNA. The model is about 6 feet high, and made of metal plates and rods. It is a scale model of a short stretch of the DNA molecule, representing the sizes and locations of its atoms, expanded to nearly a billion times actual size.

Models of this type, made of metal, cardboard, plastic, or wood, enable students and scientists to see the three-dimensional structure of molecules. They resemble and are used rather like children's construction toys, though with more serious intent. In recent years computer graphics have largely replaced physical models, which are impractical when thousands of atoms are involved, as is often the case, ✦ but even on the computer the model is a three-dimensional scale representation of the actual shape of the molecule.

Physicists and astronomers frequently talk about models, but mean something very different. For them a model is much more abstract. It is a table of numbers, some equations, or even just a few qualitative ideas and assumptions. For a comparatively well understood object such as the Sun a model consists of thousands of numbers, computed from equations that describe how the temperature, pressure, density, chemical composition, and other physical quantities vary from the center of the Sun to its surface. If the equations are correct, then the numbers are likely to be accurate; this accuracy can be tested by using them to predict observable quantities, such as the properties of vibrations of the Sun (which, amazingly, can be studied in great detail by observing its surface).

In trying to understand a new phenomenon, such as gamma-ray

bursts, an astrophysicist invents a much cruder model, often entirely qualitative. This model consists of asserting that something happens, and that this event will produce the phenomenon he is trying to explain. For example, he may suggest that a comet will fall onto a neutron star (a star roughly the mass of the Sun but only about 20 km across, and therefore possessed of an extremely strong gravitational field) and that this impact will make a gamma-ray burst. This particular model was proposed by the Cornell University astrophysicists Martin Harwit and Edwin Salpeter in 1973, and in the state of knowledge at the time deserved careful consideration. I now use it as a poster child for astrophysical modeling because it is a particularly clear illustration of how we deal with a new and inexplicable discovery.

Creating such a model sounds easy, and of no more value than a child's fantasies of flying. If all the scientist had done had been to daydream as a child does it would have been worthless. Such fantasies are occasionally published, but are not taken seriously. A more substantial model is the product of careful, though mostly qualitative, thinking. It is a hypothesis designed to explain the observations in an intellectually economical way, showing how natural processes happen to produce the phenomena we observe. Nature, without conscious intent, is more inventive and subtle than the cleverest scientist, who must follow in her path, picking up the clues she heedlessly strews.

To be useful the scientist must then try to calculate, as quantitatively as possible, the consequences of his model and to show that it can actually explain the phenomenon it purports to model. He must also demonstrate that its assumptions are reasonable. Ideally, the calculations will be well defined and feasible, and will lead to definite predictions that may be compared to extant data or to data that can be obtained from new observations or experiments. Comparison of the data to the predictions will prove the model right or wrong.

Unfortunately, the answer is rarely so clear. Occasionally the calculation itself is the obstacle, requiring computer power beyond that available. In such cases the rapid increase of computer speed and memory offers hope that the calculation impossible today will be possible soon. However, this situation is unusual, and the extraordinary increase in computer power in recent decades (more than a millionfold in the last 40 years, and perhaps a trillionfold compared to pencil-and-paper calculation) has, by itself, solved few significant scientific problems. The reason is that a determined scientist is always able to substitute ingenuity for computer power, and does so if computer power is the only necessity lacking.

More often the difficulty lies in two tougher nuts, called initial conditions and turbulence. The technical term "initial conditions" has the same meaning these words have in everyday speech: the starting point of the phenomenon being studied. This is particularly difficult to determine in astrophysics because its phenomena occur naturally, rather than being carefully arranged in the laboratory. In fact, it may be said that the chief difference between physics and astrophysics is that the task of physics is to determine the laws of nature and calculate their consequences, while the task of astrophysics is to determine how nature happened to set up the initial conditions of phenomena whose laws are known.

For example, consider the hypothesis of comets falling onto a neutron star. The scientist considering it must ask how often such events occur, to see if they are frequent enough, but not too frequent, to explain gamma-ray bursts. To answer this question requires knowing, among other things, how many neutron stars there are in the galaxy, how many comets orbit them, and what kind of orbits the comets have. All these quantities are initial conditions for this gamma-ray burst model, for its calculation must begin with them.

Astronomers have some basis for estimating the number of neutron stars because they are observed as pulsars or X-ray sources, and because we observe the supernova explosions in which some (perhaps all—this is not known) neutron stars are born. However, we have almost no basis for estimating how many comets orbit the average neutron star. There are estimates, though very uncertain ones, for the number of comets orbiting the Sun, but we know nothing about the comets orbiting other stars. Worse, the formation of a neutron star is a violent event that may, or may not, destroy the comets orbiting that star, depending on just how violent it is and on how far away the comets are.

The attempt to calculate the rate at which comets fall onto neutron stars runs aground because we do not know the essential initial conditions, and we cannot calculate them because we do not know how many comets orbit the stars that will, later in their lives, become neutron stars. We also do not know what fraction of the comets survive this process, and cannot calculate it because of similar uncertainties concerning the formation of neutron stars (Is mass expelled? How much? How fast? In all directions or only in some? etc.). We cannot remove these uncertainties because the final stages of the lives of stars depend on yet other poorly understood processes. We are stymied, and can only make half-educated guesses.

Suppose we turn away from these difficulties and simply ask what happens if a comet does fall onto a neutron star. Will the resulting event resemble a gamma-ray burst? Here, we can do a little better. We can calculate (assuming we know the mass of a "typical" comet, an initial condition that is uncertain in our solar system, and completely unknown anywhere else) how much energy will be released. This is high school physics. The result gives an upper limit to the amount of energy that can be radiated as gamma rays. It is only an upper limit because some, perhaps all, of the energy released may appear in other forms. Comparing this maximum energy to observations of gamma-ray bursts gives estimates for their distances (actually upper bounds, because if energy appears in forms other than gamma rays, fewer gamma rays are produced and their source must be closer).

When the calculation is performed, using values actually observed, the result is that a typical gamma-ray burst must be closer than about 1000 light-years, while the brightest bursts ever observed must be no more than about 10 light-years away. These results, placing the sources of gamma-ray bursts in the solar neighborhood, are reasonable; the fact that the faintest bursts are implied to be closer than the thickness of the galactic disc is consistent with the isotropic distribution of bursts on the sky, while the estimate of 10 light-years for the closest is consistent with independent estimates of the distance to the closest neutron star. For a quick response to the discovery of gamma-ray bursts, this model was doing well.

Every air traveler is familiar with turbulence in the atmosphere, especially the updrafts and downdrafts that make an airplane bounce up and down. Turbulence consists of rapid and irregular variations in the air's velocity, and other physical quantities, from place to place. Its effects range from barely perceptible motion to (rarely) tossing the airplane around like a child's toy in a giant hand, slamming meal trays and unbelted passengers into the ceiling. Pilots often announce that turbulence is expected, but only because they have been told by other pilots who passed through it a short time earlier, were warned by instruments that detect it ahead of the airplane, or are aware of the meteorological conditions that frequently make turbulence. It cannot be predicted reliably or in detail and occasionally strikes without warning.

Air passengers' discomfort is hardly the only consequence of turbulence. The flow of fluid around a body creates turbulence in its wake, even if there was none in the fluid ahead, exerting friction on flying and swimming bodies, resisting the flow of fluid through pipes, and mixing the sugar in your coffee cup and the fuel in engines. Turbulence is a nearly universal feature of fluid flow, but it is not uniform or predictable; unpredictability is the signature, even the definition, of turbulence.

If a comet falls onto a neutron star, its matter flows freely, just as a marshmallow may be crushed in your fist and squeezed out between your fingers, for its strength is insufficient to resist the enormous forces imposed by the neutron star's intense gravitational and magnetic fields. The comet is stretched into a fluid jet, like toothpaste from a tube. If it approaches the neutron star off-center, as will almost always be the case because bull's-eyes are even rarer in nature than at the dartboard, it will go into orbit around it. Exactly what happens then is incalculable, partly because the initial conditions (e.g., the neutron star's magnetic field and its orientation) are not known, but also because the comet's matter flows as a turbulent fluid. How quickly it falls from this orbit onto the neutron star, the part of this process that can release the energy necessary to make the gamma-ray burst, depends on turbulence in the orbiting matter and is a matter of guesswork.

The comets-falling-onto-neutron-stars model passed a minimal test of plausibility—with reasonable assumptions, it was shown to be consistent with the observed brightnesses of gamma-ray bursts because comparing its predictions to the observed brightnesses led to reasonable inferences for their distances. One qualitative feature looked particularly attractive: comets are likely to differ in size and shape, as the magnetic fields of neutron stars differ in strength and orientation, and comets will fall onto neutron stars in different ways (different distances off-center). All this is suggestive of the observed heterogeneity of gamma-ray bursts. It might also be possible to explain how often bursts occur; at least, our ignorance of the number of comets orbiting neutron stars made it impossible to disprove this. Unfortunately, the model could not be compared in detail to the observed data to see if it could explain the durations, complex time structure, and gamma-ray spectra of bursts because in the model all these properties depend on incalculable turbulent flows. The poster child's diseases, as well as its attractiveness, were apparent.

By the rules of theoretical astrophysics this model was promising, but it was hardly the only model considered. Ruderman's 1974 review listed dozens of models, and a more recent review counted a hundred. Many of them were close relatives, and there were only a few truly different ideas. Like comets falling onto neutron stars, most models were sufficiently energetic to explain gamma-ray bursts if they occur in the solar neighborhood, but not at much greater distances. Modelers were chiefly concerned with the temporal character of gamma-ray bursts—the fact that they were brief, sudden, unpredictable, and rare events with complex and diverse subpulse structure. None of these early models attempted to explain the enormously large energy output that would be required if bursts were at cosmological distances. Until the distance scale was definitively determined, models would be preferred on the basis of how well they explained the temporal behavior of bursts.

Models of gamma-ray bursts may be grouped in different ways. For example, there are several possible sources for the energy radiated in the burst: gravitational, thermonuclear, rotational, and magnetic.

Gravitational energy is released when matter is pulled in by the gravitational attraction of another body. It is the energy that enables a hailstone to dent a car, a heavy falling object to smash your toe, or falling water to turn a water wheel or hydroelectric turbine. Astronomers believe gravitational energy powers many X-ray stars, including the most luminous ones, in which matter from a nearby companion star falls onto a neutron star or black hole.

Thermonuclear energy powers most ordinary stars, such as the Sun. It also drives some supernova explosions and all of the lesser eruptions known as novas, as well as hydrogen bombs, so it is a natural candidate for any sort of brief transient energy release. The initial heat to ignite thermonuclear energy release in stars comes from gravitational energy.

Rotational energy powers radio pulsars, the neutron stars that radiate pulses of radio waves with regular periods ranging from about a thousandth of a second to 10 seconds. Rotational energy is, in the language of physics, kinetic energy, meaning that it is the energy of a moving object, in this case the matter of the neutron star rotating about its axis. Some gravitational energy is converted to rotational energy when an ordinary star collapses to make a neutron star.

Magnetic fields also contain energy, and on the Sun this energy is released in the form of flares, brief localized bursts of visible light, radio waves, X rays, and streams of energetic subatomic particles. These particles make auroræ and occasionally interfere with short-wave radio communications on Earth.

Another basis for sorting models of gamma-ray bursts is the nature of the astronomical objects involved. This is logically independent of the source of energy, so constructing a model becomes a matter of choosing one from column A (the kind of energy released) and one from column B (the type of object). Possible components include everything known to astronomers: black holes, neutron stars, white dwarf stars (stars with roughly the mass of the Sun but the radius of Earth), ordinary stars, planets, comets, dust grains, and a few more speculative possibilities (white holes, cosmic strings, wormholes) that probably do not exist at all. There are many possible combinations, which accounts for the large number of proposed models.

The first task of a model is to account for the energy requirements. Einstein's famous equation $E = mc^2$ established the equivalence of mass and energy, where E is an energy, m is a mass, and c is the speed of light. The factor c^2 is very large, equal to 9×10^{20} ergs per gram, where an erg is the astrophysicist's usual, and rather small, unit of energy; an erg is nearly equal to the energy required to lift a 1-milligram weight 1 centimeter in Earth's gravity, or to the energy of a million 600-KeV gamma rays. In more convenient units, 1 gram of mass is equivalent to about 20 kilotons of TNT, and 1 ounce to 600 kilotons. A respectable atomic bomb converts about a gram of mass to pure energy, and a large hydrogen bomb a matter of ounces.

Fortunately, ordinary matter cannot spontaneously convert itself to energy, so we are at no risk that the 1 ounce letter that arrived in the mail today will suddenly vaporize the city in which we live. Einstein's simple equation only sets an upper bound on the amount of energy that may actually be obtained from a given mass. It does, however, provide a convenient standard of comparison for the efficiency with which a physical process produces energy. This efficiency is a number ε between 0 and 1, the fraction of the mass m that actually is converted to energy (the rest remains as mass), so than only the energy εmc^2 is released. In the most efficient atomic or hydrogen bombs ε is less than 0.001, which is why these bombs have masses of many kilograms, rather than the tiny masses actually converted to energy.

The astrophysicist can estimate the value of ε appropriate to various models, and it is generally an advantage if ε is as large as possible, especially if gamma-ray bursts are very distant and therefore must be very energetic. The actual value of ε depends both on the physical process that releases energy and on the types of astronomical object assumed. Fortunately, reliable estimates of ε exist. For matter accreted by (falling onto) a neutron star or black hole ε is estimated to be approximately in the range 0.1 to 0.2. This implies that each gram of comet accreted by a neutron star would release about 10^{20} ergs, or about 10^{35} ergs from an assumed 10^{15} gram comet. Accretion onto a white dwarf star leads to ε of about 0.001, and accretion onto an ordinary star to ε about 10^{-6}, so it is apparent why astrophysicists considering accretion models usually assume neutron stars or black holes.

The release of thermonuclear energy gives ε that never exceeds 0.007, and is more often approximately 0.001. For a sufficiently rapidly rotating object the ε describing the release of rotational energy may be nearly as large as that for accretion, approximately 0.1–0.2 if the object is a rapidly rotating neutron star or black hole. However, rotational energy can only be extracted gradually, not in a sudden burst. Values of ε appropriate to the release of magnetic energy are very uncertain because they depend on the magnitude of the magnetic field, which can only be guessed at (the magnetic field in a hypothetical gamma-ray-burst model is at issue, not a magnetic field that has ever actually been measured); optimistic values are less than 0.001 and actual values may be 10^{-6}, 10^{-9}, or even very much less. The larger estimates of ε may be appropriate to neutron-star magnetic fields, and the smaller values to the fields of less compact objects.

This is true not only for magnetic energy but for energy release in general. When matter is compressed by gravity to make a neutron star, several forms of energy are vastly multiplied, including gravitational, rotational, and magnetic energies. This also applies to the accretional and rotational energy of a black hole (which generally has no magnetic energy). Only nuclear energy, which depends on the intrinsic properties of atomic nuclei and not on their environment, remains unchanged.

If producing the maximum amount of energy were the only criterion, all signs would point to accretion onto neutron stars or black holes. In 1974 Ruderman, recognizing these arguments, suggested "Black Hole ridden by Accretion" as the leading horse in the race among possible models of gamma-ray bursts. However, matters are not so simple. If gamma-ray bursts are comparatively close to us, then they need not be extraordinarily energetic. With their distances unknown, small but frequent eruptions would explain the data as well as giant but rare ones.

In fact, Earth receives nearly a billion times more energy, on average, in starlight (not including sunlight!) than from gamma-ray bursts. Space is filled with cosmic rays, very energetic subatomic particles (like those produced by solar flares) that also deliver nearly a billion times more power than gamma-ray bursts. The table-scraps of energy left over by other phenomena would be sufficient for gamma-ray bursts, if energy were the only criterion. Maximizing the possible energy output and efficiency may be the wrong way to choose a model. Until their distances were determined, nearly 20 years later, it was not possible to estimate the energy of an individual burst, as opposed to the average power of all bursts taken together, and averaged over time.

The second, and perhaps equally important, task of a model is to explain the observed properties other than total energy output. The most striking property of gamma-ray bursts is their complex and diverse time structure. This is unusual in astronomy. Most other eruptions, such as supernovas (the complete explosive destruction of a massive star) and novas (thermonuclear outbursts on the surface of white dwarf stars) are much simpler: their power and brightness rise rapidly and decline more slowly. In fact, the more energetic an explosion the simpler it is likely to be, just as a nuclear warhead completely vaporizes the missile that delivers it, while a conventional warhead leaves many recognizable fragments. This, in itself, suggested that gamma-ray bursts were comparatively nearby and low-energy events.

Only solar flares resemble gamma-ray bursts in their complexity and diversity. This was pointed out by Floyd Stecker and Kenneth Frost of NASA's Goddard Space Flight Center. Shortly after the discovery of bursts they published Figure 4-1, a graph of the intensity of X rays emitted by a solar flare. It lasted somewhat longer than most gamma-ray bursts, but shows two sharp intensity peaks separated by 30 seconds and a third much broader peak on a more slowly varying background. These data would fit well among the gamma-ray-burst data of Figure 1-2.

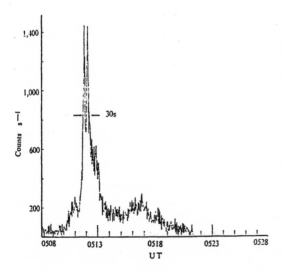

Figure 4-1. A solar X-ray burst observed in 150–250-KeV soft gamma rays, similar to the band in which gamma-ray bursts emit much of their energy. The time axis is in hours and minutes of Universal Time (roughly equivalent to Greenwich Mean Time). The spiky structure and main peak less than 2 minutes long are similar to the time history of many gamma-ray bursts (Figure 1-2). (*Nature Physical Science* V. 245, p. 71 [1973].)

Of course, gamma-ray bursts cannot be solar flares because they do not come from the Sun. However, if the Sun has flares, so likely do other stars. In most cases flares of the power observed on the Sun would be undetectable at the distances of other stars, but there are a few intrinsically faint stars whose flares are powerful enough in comparison to their steady luminosity to be readily observable (with characteristic directness, astronomers call them flare stars).

It is known for the Sun, and assumed for other stars, that flares are powered by the release of magnetic energy. The magnetic field and energy available on the Sun are comparatively small. Its general magnetic field is about 10 gauss (the usual unit in which magnetic field is measured; for comparison, Earth's field is about 0.6 gauss and the field of a child's magnet about 1000 gauss), although it can be several thousand gauss in sunspots. The magnetic energy of the Sun as a whole is about 10^{34} ergs. This would only be enough to power the Sun for a few seconds (its power actually comes from thermonuclear energy released near its center), and cannot be released fast enough to power a gamma-ray burst. Flares like those observed on the Sun are not powerful enough to explain gamma-ray bursts, even at the distances of the closest stars.

Other stars are known to have much more intense magnetic fields. Some strongly magnetized but otherwise ordinary stars have 10^{38} ergs of magnetic energy, while there are magnetic white dwarf stars with 10^{42} ergs and neutron stars with 10^{43} ergs. Comparing a flare on such a star to one on the Sun is like comparing a hydrogen bomb to a firecracker, which differ both in the amount of energy released and in how they work. Yet greater magnetic energies are conceivable; values as high as 10^{47} ergs have been suggested for certain neutron stars (unsurprisingly, termed magnetars) and may have been observed, although the evidence remains controversial. Even the highest estimates of the available magnetic energy were much less than other sources of energy, but were sufficient to provide the modest average energy requirements of gamma-ray bursts, if only the energy appeared as brief eruptions. The complex subpulse structure of solar flares then made it natural to suggest that gamma-ray bursts might be produced by analogous flares on some kind of magnetized stars.

Simply inventing a model was easy, perhaps the work of an afternoon. Making rough estimates of its possible energy production is not much harder. The real difficulty lay in deciding if the model could really make gamma-ray bursts with the observed properties. Here the usual, and usually insurmountable, obstacles of initial conditions and turbulence arose. In the case of accretion models the difficulty was both initial conditions and turbulence: would there be a suitable source of matter to accrete at the right rate, and would turbulent flow give the right variability, to produce the observed behavior?

Flare models suffered chiefly from turbulence. The process by which magnetic energy is converted into energetic particles in a flare, and then into X rays and gamma rays, depends on a beast called plasma turbulence, more fearsome even than the ordinary fluid turbulence felt by the air traveler. A plasma consists of matter in which electrons have been stripped from their atoms; all matter turns into plasma if its temperature is more than a few thousand degrees, and nearly all matter in the universe is plasma. Because the electrons in a plasma are free to move they are subject to electric and magnetic forces that do not affect other fluids. Many additional kinds of waves and turbulence can exist in a plasma in addition to the surface waves, sound waves, and eddies we see in air and water. Plasma turbulence is difficult enough in a carefully controlled laboratory experiment. In a complex natural environment, with a dash of uncertainty as to the initial conditions thrown in, the result is an incalculable mess.

After nearly a century of detailed study, even solar flares are not understood. It is unclear why they occur at all. Flares on remote stars, especially such exotic objects as white dwarfs and neutron stars, remain, as Churchill described Russia, a riddle wrapped in a mystery inside an enigma. The question of whether such superflares would occur, and what their properties would be if they did occur, could not be answered; having a hydrogen bomb at hand is irrelevant if it cannot be triggered.

It was clear that the question of choosing the right model could not be answered by pure thought alone. More astronomical data would be needed. If gamma-ray bursts could be connected to some other astronomical object, that would be a valuable clue. Alternatively, if their distance scale could be determined, that would constrain their energy scale and would limit the acceptable models. Further progress would require new data. Some years passed before these data became available, and when they did they were not always what they seemed.

5

Compactness

Ten years before physicists and electrical engineers discovered gamma-ray bursts, astronomers discovered quasars. The astronomers were using large optical telescopes to search for the origin of strong cosmic radio sources. Radio telescopes produce only fuzzy images (like a myopic person trying to read fine print), and the earliest radio telescopes, built in the 1950s, were particularly poor. It was clear that radio waves came from the Sun, Jupiter, the center of our galaxy, the remains of exploded stars (supernovas), and a few bright nearby galaxies, but the identifications of most radio sources were not apparent, leading to a mystery very similar to that of the origin of gamma-ray bursts.

In the early 1960s astronomers built larger radio telescopes that produced sharper images and located the radio sources more accurately. When optical astronomers turned their telescopes in these directions, often all they found were faint points of light looking exactly like dim stars. Naturally, these were named quasi-stellar radio sources, a term soon shortened to quasars.

To learn more about the objects they study, astronomers pass their light through a spectroscope, an instrument that breaks up the light into all its constituent colors, just as a prism or water droplets make a rainbow of sunlight. In fact, early spectroscopes used glass prisms. Modern professional spectroscopes use diffraction gratings, which consist of a series of very closely spaced (less than a 10,000th of an inch apart) grooves cut into the surface of a flat piece of glass, or stamped onto a sheet of transparent plastic or reflective foil. The same principle is used in holograms to produce vivid colors that change when the hologram is tilted. It also creates the natural colors of butterflies.

Light consists of waves of electric and magnetic fields, somewhat like waves on water. The distance between successive peaks of these waves is called their wavelength. The wavelengths of visible light are very short, ranging from 0.00007 cm for red light to 0.00004 cm for blue light. Prisms and diffraction gratings work by bending light of different wavelengths by different angles.

If a broad beam of light falls on a prism or diffraction grating, then each part of the beam is broken up into a separate rainbow of colors. These rainbows overlap, and the colors wash out. To avoid this, the light that enters a spectroscope must first pass through a narrow aperture or slit, so that only one rainbow is formed. This rainbow contains all the colors, or wavelengths, of light clearly separated into a broad band, with red at one end and blue at the other. The astronomer records this light electronically. The electronic image is not colored, but the color (or wavelength) of the light may be read off directly from its position in this broad band.

The distribution of energy in the various wavelengths of light is called its spectrum. Cooler stars have much more red light than blue in their spectra; the hottest stars have much more blue than red. The colors of a few of the brightest stars may be seen with the naked eye. An electric range or heater gets hot enough to radiate a visible red or orange glow. The filament of an incandescent light bulb is hotter, and appears yellow-white to the eye, and the Sun, hotter still, appears white.

Gamma rays are electromagnetic waves, but their wavelengths are a hundred thousand, or more, times shorter than those of visible light. The bending of gamma rays by prisms or diffraction gratings is infinitesimal, which is why their spectra are determined by the clumsy method of measuring the energy each gamma ray deposits in a detector (Chapter 2).

In the spectra of the Sun, other stars, or hot clouds of gas, the light in certain very narrow ranges of wavelength is much more intense than the light at other, nearby, wavelengths, and the light in certain other narrow ranges is less intense than nearby. These special narrow ranges of wavelength are called spectral lines because in a spectrum they appear as straight lines, bright or dark, drawn across the broad rainbow band of light. They only look like lines because they are images of the entrance slit of a spectroscope, a short, straight, and narrow line, but this name has stuck.

Astronomers learn more from the study of spectral lines than in any other way. Each atom produces its own characteristic set of lines, which makes it possible to determine the chemical composition of distant stars and galaxies (they are mostly hydrogen, the lightest element). If a source of radiation is moving, its lines are observed at frequencies different from those observed if it were at rest. This phenomenon is called the Doppler shift, and occurs for sound (listen to an approaching and receding siren) as well as for light. In the 1990s astronomers discovered planets around distant stars because the planets' gravity makes the stars move toward and away from us fast enough to produce a measurable, although very small (about one part in a million), Doppler shift in the wavelengths of their spectral lines.

When astronomers first examined the spectra of quasars they were baffled. The spectra showed lines, but these lines did not seem to correspond to the lines of any known element. Maartin Schmidt, a Dutch-American astronomer, found the astonishing explanation in 1963. The lines were those of ordinary common elements with enormous Doppler shifts. The first quasar discovered is moving away from us at 44,000 kilometers per second (15% of the speed of light), so that the wavelengths of its spectral lines are 16% longer than the wavelengths measured in the laboratory. This is why its spectrum was initially baffling: its spectral lines were shifted to unfamiliar colors where they were not expected. Only their familiar *ratios* of wavelengths, preserved because all wavelengths are multiplied by the same factor in a Doppler shift, made it possible to identify them. Most quasars are receding even faster. Some have been observed moving away from us at 95% of the speed of light, so fast that the wavelengths of their lines are multiplied more than sixfold.

Speeds as fast as quasars' were known to astronomers only in the most distant, and faintest, galaxies. The lines in their spectra are observed (first by the American astronomer Edwin Hubble in 1929) to be shifted to longer (redder) wavelengths by amounts proportional to their distances. This is called the cosmic redshift, and it shows that the universe is expanding following its birth in the Big Bang. The redshifts of the quasars were not astounding, but their brightness was. Although much too faint to be seen with the naked eye, some of them were a thousand times more luminous than entire galaxies.

Even more extraordinary was the fact that some quasars vary in brightness, as if they were variable stars. Over days, weeks, or months their brightness might double, or halve, in an irregular and unpredictable manner. This variation meant that quasars were not only, by far, the most luminous objects known in the universe, but that they were also amazingly small.

This followed from their variability by a simple argument. If a quasar contained many components, widely separated and varying independently, then the whole quasar's brightness could not vary much, because when some components were bright others would be dim. On average, these variations would cancel, at least approximately, and the quasar's brightness would be roughly steady. This was inconsistent with the observations, which showed very large variations. The quasar had to be driven by a single power source, brightening and dimming.

The size of the radiating region could be estimated from the rapidity of the variations. Consider a region powered by a varying central engine that first sends out great streams of energy, then ceases for a while, and then becomes active again, like a thundercloud lit from within by repeated bolts of lighting. A sudden and instantaneous flash sends out a wave of light in all directions, illuminating the entire cloud. But the light we see does not arrive all at once, even though all the light was emitted at the same time. Light scattered to us from the far side of the cloud has farther to travel, takes longer to reach us, and arrives later than light from the near side. This delay is not apparent to our eyes because it is only a few millionths of a second. An analogous phenomenon is observed by everyone who experiences a major earthquake. The scattering of seismic waves in Earth stretches the shaking of the ground over many seconds.

This argument was applied to quasars soon after their discovery, with remarkable conclusions. If the intensity of a quasar's light doubles or halves in a day, then the source region cannot be larger than a lightday, the distance light travels in a day. This is, of course, $\frac{1}{365}$ of a lightyear, rather large by everyday standards, but quite small to an astronomer.

Astronomers believe the radiation—radio waves and visible light—emitted by quasars is largely emitted by a process known as synchrotron radiation, so named because it was first studied in particle accelerators called synchrotrons. When charged particles, especially electrons, move through a magnetic field their paths are bent, and this bending produces radiation. Synchrotron radiation is unobservably weak if the electrons are moving slowly, but if they are moving at nearly the speed of light it becomes significant, increasing as the square of the electrons' energy. The radiated power is drawn from the energy of the electrons, which gradually slow, as if emitting radiation were a form of friction. From the observed power output of a quasar and its size, as estimated from the rapidity with which it is observed to vary, it is easy to calculate the intensity of radiation within it. Its power is high and its size is small, so the intensity must be very great.

In fact, the calculated intensity turned out to be so high that another physical process takes over. Electrons can collide with radiation. This is called Compton scattering, named for the American physicist who discovered it in 1922. If the electron is very energetic, moving at nearly the speed of light, the collision transfers energy from the electron to the radiation, just as a tennis racket or baseball bat slamming a ball gives energy to the ball.* The more radiation is present, the more likely an electron is to collide with it, and the faster it loses energy.

Energy flows from the electrons to the radiation by Compton scattering as well as by the emission of synchrotron radiation. In some quasars the radiation intensity is so high that the electrons lose more energy, and the radiation gains more energy, by Compton scattering than in synchrotron radiation itself. And not just a little more, but ten, a hundred, or a thousand times more, depending on just how the numbers work out. But now something worse happens. The Compton scattered radiation, with its multiplied energy, can suck even more energy, multiplied again by the same large factor, from the electrons in a second round of Compton scattering, and then a third round can extract even more—almost without end. This became known as the Compton catastrophe. It was pointed out in 1966 by the British and American astronomers Fred Hoyle, Geoffrey Burbidge, and Wal Sargent.

The Compton catastrophe is so rapid that the electrons lose their energy before they can fill the volume from which synchrotron radiation is observed. The model contradicts itself: it begins by asserting that the electrons from a central engine fill a radiating region of a certain size (inferred from the observed variability), but then it leads to the conclusion that the electrons lose their energy before they can do so. If the radiating region were smaller, so the electrons would not have to travel so far from their source, and to remain energetic so long, then the intensity of synchrotron radiation would be even higher (because all the observed luminosity would have to come from a smaller volume), and the inconsistency would be even worse. If the radiating region were larger, the observed variations in brightness could not be explained.

The evidence for synchrotron radiation was strong, and no other known process could explain the observed spectra of quasars. Seeing

---

*Actually, Compton observed the scattering of gamma rays by electrons at rest, in which case the radiation gives energy to the electrons. The process important in quasars is properly called inverse Compton scattering, but because the fundamental interaction is the same the "inverse" is usually dropped.

no alternative, Burbidge and Hoyle concluded that the distances (and therefore the luminosities and intensities of radiation) of quasars had been drastically overestimated. This required them to reject the interpretation of quasar redshifts as part of the general expansion of the universe, and along with that they rejected Big Bang cosmology entirely. This rapidly became a fringe position, for a great body of astronomical evidence supports the conventional interpretation of quasar redshifts and distances.

The problem was solved within months by the Dutch astronomer Ludwig Woltjer. He found a flaw in the assumptions behind the Compton catastrophe. If the radiating electrons were rushing outward, away from the central engine in which they were energized, then their synchrotron radiation would also be beamed outward. With the radiation (light and radio waves) flowing out at the speed of light, and the electrons streaming out nearly that fast (typically at 99.9999% of the speed of light), electrons and radiation would undergo Compton scattering much less often than if they were moving randomly in all directions, as had been tacitly assumed. There would be no head-on, or even broadside, collisions, only the gentle tap of the bumper of two cars on a freeway, the one behind moving very slightly faster than the one ahead. Woltjer showed that this reduced the energy transferred by Compton scattering by an enormous factor. There need be no Compton catastrophe after all.

The problem of compactness appeared again for gamma-ray bursts. Like quasars, they are very small and very luminous. In fact, in each respect they exceed quasars by large factors. They vary in tenths of seconds rather than days, roughly a million times faster than quasars, so that, by the now-familiar arguments, they must be roughly a million times smaller. The luminosity of gamma-ray bursts was unknown, as long as their distances were unknown, but it was apparent that if they were at great distances their luminosities would be enormous. If they were at cosmological distances, then, during their brief lives, they could be a million times brighter than quasars. Combined with their tiny inferred sizes, the compactness problem would return, a trillion times worse.

This was pointed out, in two separate papers in 1978, by Giacomo Cavallo and Martin Rees at Cambridge, England, and Wolfgang Schmidt at NASA. Of these, the most prominent by far was Rees. He began his career in the mid-1960s, when the discovery of quasars revived and energized astronomy. This was nowhere more true than in Britain, where theoretical astrophysics had been in hibernation for a generation. Rees's first work was on the theory of quasars, including the supposed Compton catastrophe. Obviously very talented, he quickly came to dominate British astrophysics, casting such a shadow that many younger scientists left for America to escape it. He was soon a fixture on the international conference circuit, giving invited talks at, seemingly, every meeting on quasars, cosmology, and, later, gamma-ray bursts. In his bachelor days he had something of a reputation as a ladies' man, despite a prominent spinal deformity resulting from misdiagnosed and mistreated childhood scoliosis. Remarkably young, he became the Plumian Professor of Astronomy and Natural Philosophy at Cambridge, the most prestigious astronomical position in Britain. Later, he was knighted and appointed Astronomer Royal. Although his name and heritage are Welsh, this quintessential member of the British establishment does not consider himself Welsh, which he even spells with a lowercase w, as if it were not a proper nationality. A childless only child, his science would be his legacy.

A great deal of the theoretical work on quasars would later be applicable to gamma-ray bursts, because both are small, energetic, and rapidly varying. The first instance of this was the compactness problem. It took a different and more general form for gamma-ray bursts. There was no direct evidence that synchrotron radiation was the source of the emitted radiation of bursts, but Cavallo, Rees, and Schmidt discovered a compactness paradox based only on the fact that gamma-ray bursts emit gamma rays. Einstein's equation $E = mc^2$ gives the amount of energy E that can be obtained if a mass m is completely turned into energy. This relation can be turned around: if two gamma rays with total energy E collide, they may produce a mass m.

However, this is only possible if particles whose masses are m or less can be created (visible light cannot turn into matter because there are no particles with small enough masses). The least massive known particles are electrons (negatively charged) and positrons (positively charged), each with a mass corresponding to 0.511 MeV of energy. Because electric charge is never created or destroyed, electrons and positrons can only be created in pairs, one of each, with zero total charge. Two gamma rays, each of energy 0.511 MeV or more, colliding head-on, can therefore produce an electron–positron pair. If the collision is not head-on, then the necessary energy is greater. If the gamma rays have more energy than the minimum required, the extra appears as kinetic energy of the newborn matter—the electron and positron are born in motion.

Cavallo, Rees, and Schmidt noticed that if gamma-ray bursts are very distant and luminous, their gamma rays would be so tightly packed into a small volume that they would interfere with each other's escape, as a crowd fleeing a theater fire trips over each other's feet in panic. The severity of the problem is proportional to the ratio of the luminosity to the radius. Astronomers defined a "compactness parameter" to measure it. If this number is much larger than one, the more energetic gamma rays collide with each other and produce electron–positron pairs before they can escape.

It was easy to evaluate the compactness parameter using data observed for gamma-ray bursts. If the brighter bursts were more than a few thousand light-years away, their luminosities must be so high that their compactness parameters would exceed one. Their spectrum of gamma rays would be expected to drop precipitously above 0.511 MeV, where the gamma rays had enough energy to destroy each other by producing electron–positron pairs.

No such deficiency of the more energetic gamma rays could be found in the data. The conclusion was that gamma-ray bursts were comparatively close to us, in the solar neighborhood. The dimmer bursts could be several times more distant, but still well within our galaxy.

Just as for quasars, there was a loophole in the compactness argument. In fact, it was nearly the same loophole. If the gamma rays were directed almost exactly outward, the minimum energy, called the threshold, they must have to create an electron–positron pair would be much greater, just as a collision between two cars traveling nearly parallel and at nearly the same speed is much less violent than a head-on collision, even if the cars are large, heavy, and fast. This is shown in Figure 5-1. The gamma-ray energy at which the spectrum was expected to drop could be much higher, outside the range of observation.

This was not just an arbitrary assumption. Whatever makes a gamma-ray burst releases a great deal of energy in a small volume in a short time. This energy, in some form, must stream outward along rays like the spines of a sea urchin. Adjacent spines, if they touch at all, are nearly parallel and only graze each other. Estimates of the interaction among the gamma rays, such as the compactness argument, which assume head-on collisions, cannot be accurate and may be entirely wrong.

In addition, there was a subtle error in the argument used to set an upper bound on the size of a gamma-ray burst. That argument assumed the source was at rest. If, however, the matter were moving outward at nearly the speed of light, then the actual size of the region emitting the radiation could be very much larger. Such rapidly mov-

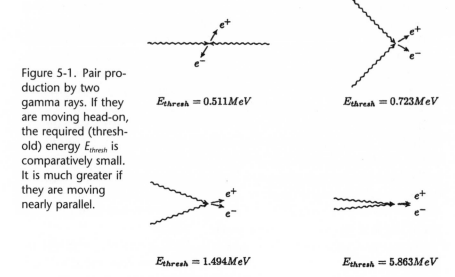

Figure 5-1. Pair production by two gamma rays. If they are moving head-on, the required (threshold) energy E_{thresh} is comparatively small. It is much greater if they are moving nearly parallel.

$E_{thresh} = 0.511 MeV$

$E_{thresh} = 0.723 MeV$

$E_{thresh} = 1.494 MeV$

$E_{thresh} = 5.863 MeV$

ing matter beams its radiation into a narrow cone in its direction of motion, so an observer sees only that tiny portion of the entire source that is moving almost exactly toward him. Any bound he sets on the size of the source only applies to this portion, and the complete source may be very much larger. In addition, the radiation is moving only a very little faster than the matter, and only slowly overtakes it. As a result, the simple argument is inapplicable to rapidly expanding sources of radiation. It was straightforward to extend the argument to sources expanding at nearly the speed of light. Any luminosity is possible, if only the matter is moving fast enough.

These subtleties were understood in the late 1970s because they had been used to understand the quasar compactness problem more than a decade earlier. They had been the subject of one of Martin Rees's first papers in 1967. In fact, they were not entirely new even then, for the theory of the scattering of a flash of radiation by a surrounding cloud (like a lightning bolt in a thundercloud, or energy from a central source in a quasar or a gamma-ray burst) had been published in detail in 1939 by a French astronomer named Paul Couderc. He was interested in explaining rings of light observed in 1901–02 surrounding a nova (a dim white dwarf star that suddenly, for a few days, weeks, or months, brightens to tens of thousands times the brightness of the Sun). These rings seemed to be expanding at several times the speed of light, even though the theory of relativity demonstrates that no object, particle, or wave can travel superluminally (faster than light)!

Couderc found the explanation. Soon after the outburst we can only see light that travels to us on an almost, but not quite, direct path from the nova. This light is scattered by dust near, but not quite on, a direct line to us, and we see it as a narrow ring around the nova. Because its path is not quite direct, it takes a little longer to arrive, and we see it a little later. As time goes on we see light that has deviated further from a direct path, arrives later, and forms larger rings (Figure 5-2). The rings, called light echoes, are parts of a much larger sphere of light from the nova, but because they are not real moving physical objects they may expand faster than light travels. The actual radiation-filled volume is much larger than the observed bright region (the light echo), so that its radiation is much more dilute than a naïve calculation would indicate and paradoxes such as the Compton catastrophe and the compactness problem need not occur.

Astronomers generally chose to neglect these possible loopholes in the compactness arguments against high luminosities and great distances for gamma-ray bursts. Couderc, originally a lycée teacher and later involved in science administration, was chiefly a popularizer of astronomy. Even more than Usov, he was essentially unknown outside his homeland. His paper, published (in French) immediately before the Second World War, was neglected and ignored until light echoes from the bright supernova of 1987 (the first supernova since 1604 observable with the naked eye) were discovered. Rees's rediscovery of Couderc's results a generation afterward, although famed for explaining superluminal expansion in quasars, was not widely applied to gamma-ray bursts. The probable reason was that other evidence pointed toward models of gamma-ray bursts as nearby, comparatively feeble, phenomena. The compactness argument, in its simplest form, seemed only to buttress that evidence. Why confuse a neat explanation by looking for loopholes and exceptions?

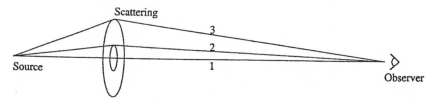

Figure 5-2. Light (or gamma rays) traveling along a straight line (1) arrives first. Light scattered along the way follows a longer path and arrives later, (2) before (3). The scattering matter, lit up like a cloud at sunset, appears as an expanding ring. Its *apparent* speed of expansion may be many times the speed of light, although nothing, not even light, is moving that fast.

6

The Large Magellanic Cloud

In 1519 the Portuguese sea captain Fernão de Magalhães, whom we call Ferdinand Magellan, commanding a fleet of five small ships, sailed under the flag of Spain to circumnavigate the globe. His purpose was to expand the Spanish empire. A papal bull of 1493 had divided the New World according to a line of longitude running through eastern South America; everything east of this line would be Portuguese and everything west would be Spanish. We see the results of this bull even today; the eastern most part of South America is Brazil, whose language is Portuguese, while the rest of Latin America speaks Spanish.

More important in 1519 were the Spice Islands, chiefly in modern Malaysia and Indonesia, the source for a valuable trade in exotic spices; the spice trade was as important in the sixteenth century as the oil trade of the Persian Gulf is today. The Portuguese had developed this trade in the decades before and after 1500 by sailing east around the Cape of Good Hope at the southern tip of Africa, and Magellan had fought in battles in which the Portuguese wrested control of the spice trade from the Muslim seafarers of the Indian Ocean. Now the Spanish wanted it. If they could prove the Spice Islands lay west of the line of the papal bull, the Islands would be Spanish. So the Spaniards hoped. When Magellan, an experienced captain dismissed and embittered by the Portuguese king, proposed to reach the Spice Islands by sailing *west* around the southern tip of South America, the Spanish court gave him ships and men. The Portuguese never forgave him.

Magellan and his crew had little interest in scientific observation and discovery for its own sake, but sailors were then learning to navigate by the stars, as amateur sailors still do, and it was important to observe the sky closely. At sea there is not much else to look at, and in the middle of a dark ocean, on a ship lit only by a few dim oil lamps, the stars appear brighter than a modern city-dweller can imagine. As Magellan's ships sailed ever farther south, looking for a passage around or through South America, the familiar northern stars disappeared behind the northern horizon and new southern stars came into view.

In 1520 an unknown member of Magellan's crew first noticed two fuzzy patches of light in the southern sky. These resembled the diffuse glow of the Milky Way, so-called because of its similarity to a band of spilled milk, which consists of billions of distant stars, each too faint to see individually with the naked eye, making up the galaxy in which we live.

The fuzzy patches discovered in Magellan's voyage are now called the Large and Small Magellanic Clouds (LMC and SMC). Like the Milky Way, they are made up of large numbers of individual stars, easily seen with a telescope. The Magellanic Clouds are small satellite galaxies to our Milky Way. They have been studied by astronomers in great detail because they are the closest examples of other galaxies, because they contain large numbers of young stars, and because they are not cloaked behind interstellar soot and dust, as is most of the Milky Way.

Stellar births are frequent in the Magellanic Clouds, and so are stellar deaths. This may seem surprising, or even contradictory, but some stars have brief lifetimes (by astronomical standards) of only a few million years; on a cosmic calendar measured in billions of years they die almost as soon as they are born. Regions of frequent stellar bursts are generally also regions of frequent stellar deaths. The famous supernova 1987A, the first supernova bright enough to be seen with the naked eye since 1604, occurred in the Large Magellanic Cloud.

On March 5, 1979, detectors aboard a record-setting nine different spacecraft observed the most intense gamma-ray burst ever recorded, before or since. The original report, made by astronomers from Los Alamos, NASA, France, and the Soviet Union, was published May 11, 1979, fast by the standards of science, especially in that pre-Internet and pre-e-mail age. In this one-paragraph preliminary announcement, known as International Astronomical Union Circular No. 3356 (distributed as a telegram or postcard), they reported its unprecedented intensity (more than ten times brighter than any recorded before), amazingly rapid onset (its intensity rose from nothing to its peak in less than a thousandth of a second), fast decline (the intensity dropped to a small fraction of its peak in a tenth of a second, but continued to be observable for a few minutes), and unusual spectrum. Comparison of the times of arrival of the gamma rays at spacecraft spread throughout the inner solar system soon showed that they came from the Large Magellanic Cloud, and not just from any location in the Cloud, but from within a supernova remnant (see Figure 6-1). A supernova remnant is the bubble of hot gas and energetic particles left behind after a massive star explodes at the end of its life; it is the astronomical equivalent of the mushroom cloud that follows a nuclear explosion in Earth's atmosphere, and typically is observable until it is about 30,000 years old. The burst of March 5, 1979, was the first gamma-ray burst to be identified with any other astronomical object.

This unexpected discovery seemed to offer several keys to the mystery of gamma-ray bursts. First, it appeared to answer the question of where they come from—from supernova remnants comparatively close to us, at least by astronomical standards. The LMC, and everything in it, is about 200,000 light-years away, while the universe is about 14,000,000,000 light-years across; in terms of cosmology, the LMC is just next door.

Second, it appeared to give strong clues as to what sort of astro-

Figure 6-1. The sky around the position of the event of March 5, 1979. The dashed line is the boundary of its possible positions as determined by an interplanetary network of detectors (further analysis shrank this error box somewhat). The solid lines are contours of X-ray intensity of the supernova remnants N49 and (N49). The axes are labeled in astronomers' coordinates of right ascension (equivalent to longitude, but measured in hours, minutes, and seconds, reflecting Earth's rotation) and declination (equivalent to latitude, with negative values south of the celestial equator). (Reprinted from *Comments on Astrophysics* V. 9, p. 17 [1980] © OPA N.V. with permission from Gordon and Breach Publishers.)

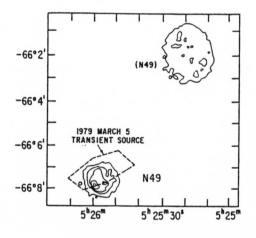

nomical object makes gamma-ray bursts. Many supernova remnants are known to harbor neutron stars, bizarre objects roughly 20 km in diameter containing about 1.4 times the mass of the entire Sun. To pack so much mass in so little volume they must be about 10^{14} times the density of water (a teaspoonful would weigh half a billion tons), and their magnetic fields are enormous. Such an object has a great deal of energy stored in its magnetic field, and can release even more energy (up to 20% of the theoretical limit $E = mc^2$) if an additional mass m were to fall onto it. A single gram (1/28 ounce) dropped onto a neutron star releases about the same energy as a 5-kiloton atomic bomb. Magnetic neutron stars were a theorist's playground for trying to understand gamma-ray bursts, and this identification of the burst of March 5, 1979, set them to work with renewed vigor.

Additional clues came from the time dependence of the gamma-ray signal itself (Figure 6-2). Its intensity rose from zero (the detector still registered some signals from other sources of radiation; this is called background) to nearly its peak value in no more than 0.0002 second. Because no signal can travel faster than the speed of light, this observation was used to argue that the size of the radiation

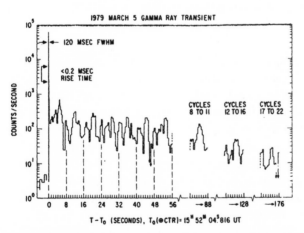

Figure 6-2. The time history of the intensity of the event of March 5, 1979. There was an initial spike of intensity lasting about a tenth of a second (more quantitatively, 120 milliseconds "full width at half maximum," the time between when it rose to half its maximum intensity and when it fell to half its maximum). This spike began abruptly, rising to near its peak in 0.2 milliseconds or less (the data were sufficient only to set this bound on its rise, not to measure it). Following the spike, radiation was measured for about 3 minutes, with gradually falling intensity, but with secondary maxima and minima repeating with a period of 8 seconds. (Reprinted from *Comments on Astrophysics* V. 9, p. 14 [1980] © OPA N.V. with permission from Gordon and Breach Publishers.)

source, all of which must have turned on in 0.0002 second or less, was no larger than the distance light travels in this time, 60 km. Even if all parts of a radiating object start emitting simultaneously, radiation from its front would arrive sooner than radiation from its back, so that it would appear to turn on gradually. The time difference would be the time required for light (gamma rays in this case) to cross the object, so that a source that turns on rapidly must be small. This fitted nicely with the idea that gamma-ray bursts are produced by neutron stars with diameters of about 20 km. This argument is not ironclad—it had been known since about 1900, and applied to astronomical objects since Couderc's paper of 1939, that much larger radiation sources could appear to turn on suddenly if they were expanding at nearly the speed of light, essentially because significant intensity would only be observed from the small portion of the source moving almost exactly toward the observer—but it all seemed to fit together with a simple picture of energy release around a neutron star.

Another important observation concerned the gradual fading of the signal. This was not a steady decline, but rather oscillated regularly: up, down, up, down, with a peak every 8 seconds (Figure 6-2). This had a natural explanation if the radiation was produced by something that rotates around an axis once every 8 seconds, and whose radiation is beamed like that of a lighthouse: first you see a bright beam, then a dimmer glow from the side of the lantern, then the bright beam again 8 seconds later, and so on, over and over. This idea is obviously sensible, and many neutron stars, called pulsars, are known to do just that, with observed rotation periods ranging from about 0.0015 to 1000 seconds. The natural, and almost certainly correct, conclusion was that the gamma-ray burst of March 5, 1979, was produced by a neutron star rotating with a period of 8 seconds. Further, it probably had a magnetic field of at least 10^{12} gauss, and perhaps much greater, because such large fields would be required to form the radiation into a beam. Pulsar fields are known, based on indirect but reliable measurements, to range from about 10^8 to 5×10^{13} gauss. The larger fields are found in pulsars with spin periods of a second or longer because the fields act as a cosmic brake pad by radiating the energy of rotation. The field estimated from the properties of the beam agreed nicely with the field suggested by the slow rotation.

In fact, it was possible to say even more. No one knew what period pulsars possessed when they were born. However, if they were assumed to be born spinning several times faster than when they were observed (just how many times faster did not matter significantly, as long as it was by a factor of two or more), and if their age could be estimated, then, using a simple and familiar theory based on elementary physics, their magnetic fields could be calculated. The first assumption was dubious (in the few cases in which it could be tested directly, it failed), but it was possible to make a reasonable estimate of the age of the supernova remnant in which the March 5, 1979, burst was found. So, although the foundations were uncertain, a tentative value of the magnetic field of the neutron star responsible for the March 5, 1979, burst could easily be calculated. Many astrophysicists appear to have done this, but apparently all thought it so obvious, or the first assumption so uncertain, that the numerical result was not actually published for many years. The answer was unprecedentedly large: about 6×10^{14} gauss.

Because the burst of March 5, 1979, was so bright, several instruments were able to measure the spectrum of its gamma rays: that is, how many gamma rays of each energy were observed. The spectrum of the burst of March 5, 1979, is shown in Figure 6-3, along with that of a typical burst for comparison. The trained eye draws several interesting conclusions from these data. The first conclusion was that during the initial spike in intensity, only about a tenth of a second long, the spectrum (i) *roughly* resembled that of a typical gamma-ray burst(iv). The actual data (i) look like an irregular staircase, rather than

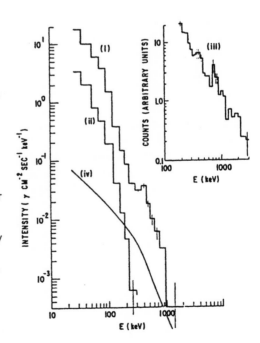

Figure 6-3. The energy spectra of the event of March 5, 1979. The data (i) are taken from its brief initial spike and (ii) from its 3 minutes of gradually decaying intensity. A gamma-ray burst of November 19, 1978, is shown (iii) in the inset for comparison, while (iv) is a smooth approximation to typical gamma-ray-burst spectra. Some evidence for a peak at about 400 KeV appears in (i), while (iii) has evidence for peaks around 400 KeV and 800 KeV; the reality of these peaks is controversial. (Reprinted from *Comments on Astrophysics* V. 9, p. 15 [1980] © OPA N.V. with permission from Gordon and Breach Publishers.)

a smooth curve, partly because the data must be divided into energy bins, just as when a demographer computes a distribution of family income he may group families with incomes between $20,000 and $30,000 into one bin, those between $30,000 and $40,000 into another bin, and so forth, and partly because of the effect of "counting statistics" (Chapter 1). The curve (iv) is smooth because it is a smooth spectrum inferred from the data, rather than actual data, such as (iii) in the inset that shows the well-studied gamma-ray burst of November 19, 1978. If you look more closely, you see that the resemblance between (i) and (iv) is not very close—(iv) has a pronounced change in steepness that is not evident in (i), and (iv) is less steep than (i) to the left of this break. However, (i) bears a close resemblance to (iii). It has long been proverbial among gamma-ray-burst scientists that bursts are all different, so that it was natural to conclude that the March 5, 1979, burst—or, at least, its first tenth of a second—was an ordinary burst.

The second conclusion rested on a comparison of curves (i) and (iii). There appeared to be upward bumps interrupting their smooth plunge. In (i) (March 5, 1979) there is a single bump at an energy of 400–500 KeV, while in (iii) (November 19, 1978) there are two such bumps, one again at 400–500 KeV and the other at approximately twice that energy. These energies are specially significant to a physicist. We saw in the previous chapter that when two gamma rays interact, as is likely to occur when they are present in enormous numbers, they will create a pair of elementary particles—the familiar electron and its anti-particle, called a positron because it is positively charged. The fate of a positron is always to find an electron with which to perform a lovers' leap of mutual annihilation, yielding two (rarely, three) gamma rays. The total energy of these gamma rays is the sum of the energies of the electron and the positron. If these particles are not moving close to the speed of light, this energy is the sum of their rest mass energies (according to the famous equation $E = mc^2$), or 1.022 MeV, shared equally between the two gamma rays, giving each an energy of 511 KeV. This "magic" number is as instantly recognized by physicists and astrophysicists as the year 1776 is by Americans, because it is observed whenever positrons are created and annihilated, whether in the laboratory or the cosmos.

The little bumps on curves (i) and (iii) occur at energies just slightly below 511 KeV. However, the astrophysicist expects this small difference, because if gamma rays are produced near the surface of a neutron star they will lose 10 to 20% of their energy in escaping from its strong gravitational field, just as a rocket launched upward gradually slows once its engine shuts off; this effect is called the gravitational redshift. Allowing for this energy loss brought the observed and measured bumps in the gamma-ray spectrum into as close agreement as anyone had the right to expect.

Even the second bump in (iii), at an energy of 800–900 KeV, had an energy consistent, allowing for gravitational redshift, with the entire electron–positron annihilation energy of 1.022 MeV. Under ordinary circumstances a positron and an electron never annihilate into a single gamma ray, for fundamental reasons derived from the theory of special relativity, but in a magnetic field exceeding 10^{13} gauss, annihilation produces a single gamma ray rather than two. Neutron stars are frequently observed to have fields approaching this value, so the entire picture—electron–positron annihilation, gravitational redshift, and strong magnetic field—hung together rather neatly, and was consistent with the 8-second inferred rotational period.

The third conclusion was a little more disturbing. The spectrum (ii)—which represents all 3 minutes of the March 5, 1979, event with the exception of the initial quarter second—was very different from that of any other gamma-ray burst. To see this it is necessary to look a little closely at Figure 6-3. All the spectra descend steeply at higher energies, but (ii) drops precipitously between 100 KeV and 300 KeV and shows *no* observable signal above 300 KeV, features not found in any of the other spectra. Spectra (i), (iii), and (iv) and spectra of nearly all gamma-ray bursts contain the majority of their energy around or above 300 KeV. This is partly hidden by the customary mode of presentation, because higher-energy gamma rays carry their energy in bigger packages (more energy per gamma ray), but only the number of gamma rays, not the energy carried, is shown. In addition, their energy is spread over a wider swath of the gamma-ray spectrum; the logarithmic (stretched) axis conceals the fact that the range 300–3000 KeV is more than ten times as wide as the range 100–300 KeV. However, no one expects all the parts of a complicated puzzle to fit together immediately, so few lost sleep over this conclusion.

The March 5, 1979, outburst seemed to point to the answer to the gamma-ray-burst mystery: they are rotating neutron stars. This was comforting, because, although they seem exotic to the layman, rotating magnetic neutron stars were the accepted explanation of pulsars, the pulsating radio beacons in the sky discovered in 1967, and binary X-ray pulsars, discovered in 1970. These had been explained to everyone's satisfaction as neutron stars, emitting radio pulses if isolated in the deep vacuum of space, but strong X-ray sources if close enough to a companion star to suck matter off its surface. Gamma-ray bursts seemed to fit nicely into this picture.

It remained mysterious why some neutron stars should emit enormous, but rare, gamma-ray bursts, while others, the pulsars, simply emit a steady stream of weak but regular pulses. In 1982 it was suggested that the neutron stars that produce bursts have magnetic fields much larger than those of pulsars, and that the bursts are giant magnetic storms (analogous to the solar flares that disrupt communications during maxima of the 11-year solar cycle, as in 1989 and 2000). This idea was strengthened by the observation, later in 1979, of additional smaller outbursts from the direction in the LMC of the great burst of March 5, 1979; like the Sun, an erupting magnetic neutron star might be expected to erupt again and again, although why and when an outburst would be triggered remained a mystery. The extraordinarily large fields calculated from the age of the supernova remnant and the neutron star's 8-second period were consistent with the idea that the outbursts were magnetic storms, powered by energy stored in the magnetic field.

A few flies remained stuck in this ointment, buzzing annoyingly, if not loudly. The first involved the locations of gamma-ray bursts. If they are powered by magnetic neutron stars, then they should be found in the same places other magnetic neutron stars are found, which is in a broad flat pancake known as the disc of our galaxy. This is the Milky Way we can see with the naked eye, and it forms a band around the sky. Pulsars have a similar distribution, but even in 1979 it was recognized that gamma-ray bursts do not. The best explanation was that we could only observe gamma-ray bursts out to comparatively short distances, less than the thickness of the pancake (about 300 light-years), so that they would appear with approximately equal likelihood on all directions. This is the reason the stars close enough to be seen with the naked eye are not concentrated into the band of the Milky Way. However, it was hard to reconcile this with the fact that the March 5, 1979, outburst in the LMC, about five hundred times more distant than the thickness of this pancake, was also the brightest gamma-ray burst. The most distant object of a class should be the faintest, not the brightest.

More buzzing arose from the distinctive shapes of the spectrum and the time dependence of the March 5, 1979, burst. Its spectrum (ii) (Figure 6-3) was quite different from the typical gamma-ray-burst spectrum (iv). In addition, no other burst showed any of the characteristic temporal features of March 5, 1979: the submillisecond rise, the initial tenth-of-a-second spike, the steady but gradual decay, and the periodic variation in intensity during this decay. In 1979 these differences were usually dismissed with the remark "They're all different."

This may have been justified at the time, although the proverbial "When you've seen one gamma-ray burst, you've seen one gamma-ray burst" did not imply that gamma-ray bursts had *any* conceivable time history, but only that they were all different. It is difficult to quantify the enormous range of possible intensity versus time graphs, but that did not justify ignoring the plain qualitative distinctions that could be made between March 5, 1979, and all other bursts. In addition, closer examination later showed that the spectra of gamma-ray bursts other than March 5, 1979, were all remarkably similar, and, in fact, an important distinction was overlooked. March 5, 1979, and other gamma-ray bursts really had only two things in common: observation by the same instruments in approximately the same region of the electromagnetic spectrum, and durations of a few minutes.

The history of exploration of new astronomical phenomena, especially those observed by nontraditional methods such as radio, X-ray, and gamma-ray instruments, has usually involved determining accurate positions on the sky, followed by identifying the new phenomenon with something that could be studied by powerful traditional means, such as optical telescopes. Astronomers greeted the burst of March 5, 1979, with joy and excitement because its accurate coordinates led to an immediate identification with a previously studied object, a supernova remnant. This blinded them to the possibility that they had not really solved the problem they thought they had solved, somewhat like the apocryphal drunk who looks for his keys under the street light because that's where the light is. This drunk found someone else's keys.

7

False Lines

The power of optical astronomy is in spectroscopy. By breaking light up into its component colors in a spectroscope, astronomers determine the motions, composition, temperature, pressure, magnetic field, rotation, and pulsation of the stars they study. Gamma-ray and X-ray astronomers (the gamma-ray and X-ray bands overlap, so gamma-ray and X-ray astronomers often face similar instrumental problems) would like to do the same, but they find spectroscopy much harder than it is for visible light.

There are several reasons for this. One is instrumental. Prisms do not bend X rays or gamma rays enough to be useful. To disperse these radiations in a spectroscope a diffraction grating must be used. The gratings used in visible light spectroscopy are inadequate for gamma rays because their wavelengths are tens of thousands of times shorter than those of visible light, although these gratings can be used for the lowest-energy X rays. Instead, the atoms in a crystal can be used as a diffraction grating for higher-energy X rays and gamma rays, a fact, discovered by the father–son team of William and Lawrence Bragg in 1912, which helped demonstrate that X rays are electromagnetic radiation and that all matter is made of atoms (and which made Lawrence Bragg, at 25, the youngest Nobel prize winner ever).

Unfortunately, once dispersed by a diffraction grating X rays or gamma rays are very weak, and can be detected only if their source is intense or if a very long exposure is made. This is fine in the laboratory, where X-ray diffraction is a widely used tool. It is also used in the astronomy of very-low-energy X rays, which can be focused by large X-ray telescopes, concentrating them enough that they are detectable even after dispersion. It does not work for higher-energy X rays or gamma rays that cannot be focused. It fails twice over for gamma-ray bursts because of their unpredictability: even if the gamma rays could be focused, the burst would be over before a telescope could be pointed in its direction.

A second difficulty is that ordinary atoms produce very many spectral lines in visible light and low-energy X rays, but few in higher energy X rays and none in gamma rays. Stars are made up almost entirely of light atoms. They are about 70%, by mass, hydrogen, the lightest, about 28% helium, the next lightest, about 2% carbon, nitrogen and oxygen, light elements that are the building blocks of life, a few tenths of a percent of light metals such as calcium, magnesium, and aluminum and middle-weight iron, and only tiny quantities of elements heavier than iron. Iron has a spectral line at an X-ray energy of about 7 KeV, well below the gamma-ray band, and lighter elements have all their X-ray spectral lines at even lower energies. Gamma-ray-burst astronomers therefore generally study only the broad distribution of radiation across the spectrum, rather than atomic spectral lines. For this they use detectors, such as NaI(Tl) scintillators, which provide only a rough measure of the energy of each gamma ray, but do not require that the radiation, already weak, be further diluted by spreading it into a gamma-ray rainbow.

If a gamma-ray line is strong, it may be detected even with a NaI(Tl) scintillator, the instrument used in most gamma-ray-burst studies. In fact, before the invention of germanium detectors nuclear physicists used NaI(Tl) scintillators to study gamma-ray lines from atomic nuclei. This is much easier for the nuclear physicist than for a gamma-ray-burst astronomer because laboratory experiments are more controllable than unpredictable astronomical events and because nuclei radiate their energy in a few well-separated gamma-ray lines, rather than spreading it across the entire gamma-ray spectrum as a burst does. If the line is strong enough, an excess (or deficiency) of gamma rays in a narrow range of photon energy, the definition of a spectral line, may stand out in the spectrum, even though the scintillator gives only a rough measure of the energy of each gamma ray detected, confusing many of the gamma rays in the spectral line with those at somewhat greater or lesser energies. This is possible only if there are spectral lines in the energy range observed, typically from about 30 KeV to 3 MeV. Atoms (except for the very rare heavy ones) have no spectral lines in this range.

Fortunately, there is another physical process that can, in principle, produce spectral lines in any part of the electromagnetic spectrum. This is called cyclotron radiation, because it occurs in cyclotrons, a type of particle accelerator. Magnetic fields exert forces on moving charged particles, guiding them on circular (or helical) orbits. As a particle traces out this path, it produces electromagnetic radiation whose frequency is the same as the frequency with which it goes around its circular (or helical) orbit, just as a tuning fork vibrating 256 times per second radiates sound waves with a frequency of 256 cycles per second (middle C). Cyclotron radiation is closely related to synchrotron radiation (Chapter 5): if a charged particle is moving at nearly the speed of light, its radiation is called synchrotron radiation (and occurs at much higher frequencies than the cyclotron orbital frequency), while if the particle moves much more slowly than the speed of light it radiates cyclotron radiation.

The frequency of cyclotron radiation is proportional to the magnetic field, and inversely proportional to the mass of the charged particle. The power radiated is proportional to the square of the magnetic field and inversely proportional to the cube of the mass of the particle (of a given energy). Therefore, cyclotron radiation is important if the magnetic field is large and for light charged particles (electrons and positrons). It is generally insignificant for small fields and heavier particles.

Soon after pulsars were discovered in 1968 it was realized that they could be explained as rotating magnetized neutron stars. The spin of such a star will gradually slow down because it loses energy by radiating an electromagnetic wave at its frequency of rotation, in rough analogy to cyclotron radiation by a charged particle at the frequency of its motion in a magnetic field. This radiation by a neutron star is at a very low frequency (from less than one cycle per second to about 1000 cycles per second), and cannot be observed directly because it is absorbed by interstellar gases. However, the slowing of pulsars' spin was predicted theoretically and was soon observed, confirming that they really are rotating magnetized neutron stars. The strength of the magnetic field may be calculated directly from the rate at which the spin slows. The results are scattered over a very broad range, but typical pulsar magnetic fields are between 10^{11} and 10^{13} gauss.

In fields of this strength electron cyclotron radiation is extremely powerful and takes the form of X rays or gamma rays. X-ray astronomers looked for it first, in an X-ray pulsar called Hercules X-1 (the first X-ray source to be discovered in the constellation Hercules). This object had been studied and analyzed in remarkable detail. It contains a neutron star rotating once every 1.24 seconds, sweeping an X-ray beam around with its rotation (in fact, the lighthouse model for the 8-second period of the gamma-ray burst of March 5, 1979, was borrowed directly from work on Hercules X-1). The neutron star orbits an ordinary star somewhat more massive and several times more luminous than the Sun.

In 1976 a group of X-ray astronomers, under the leadership of Joachim Trümper in Germany, flew a NaI(Tl) detector on a balloon to search for a cyclotron line in the spectrum of Hercules X-1. The detector was the same type used in most gamma-ray-burst instruments, although designed to study photons of energy from about 15 KeV to 135 KeV. This spectral region, lower in energy than that used in gamma-ray-burst satellites, has been called, at different times, either X rays or gamma rays.

The results were spectacular. As shown in Figure 7-1, there is a strong emission line at a photon energy of 58 KeV, corresponding to electron cyclotron radiation in a magnetic field of about 5×10^{12} gauss. This was a remarkable confirmation of expectations. The magnetic field was, by a large factor, the most intense ever measured directly (rather than inferred from a spin-down rate). The spectrum also shows evidence for a second emission line around 110 KeV, nearly twice the energy of the first line. Such a "second harmonic" line is

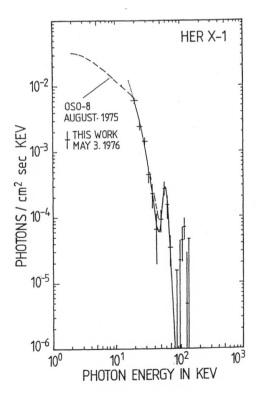

Figure 7-1. The X-ray spectrum of Hercules X-1. The error bars are the actual measurements, showing (vertical extent) statistical uncertainties and (horizontal extent) the range in energy averaged into each data point. Solid, dashed, and dotted curves show various mathematical expressions fitted to data, including an earlier experiment on the satellite OSO-8. (Reprinted by permission from *Astrophys. J. Lett.* V. 219, p. L109 © 1978 American Astronomical Society.)

analogous to the overtones produced by musical instruments. It was also predicted by theory.

This success inspired gamma-ray-burst astronomers to search for similar cyclotron lines in the spectra of bursts, particularly after the giant burst of March 5, 1979, revived interest in bursts and strengthened the arguments for their being magnetic neutron stars. In 1981 a Soviet group under the leadership of Evgeni Mazets published a paper reporting lines in the spectra of nearly thirty gamma-ray bursts. In some bursts they found emission lines at energies of about 400–450 KeV, consistent with electron–positron pair annihilation near the surface of a neutron star. This was similar to what had been reported for the burst of March 5, 1979. In many more bursts they reported absorption lines (a deficiency of gamma rays in a narrow range of energy) between 40 and 70 KeV, similar to the energy of the emission line found in Hercules X-1. Some of these data are shown in Figure 7-2. Seven years later an independent Japanese experiment found similar results in two other gamma-ray bursts.

Astronomers were not disturbed by the fact that most of the lines are in absorption rather than emission. A physical process, such as cyclotron radiation or the motion of an electron in an atom, which produces radiation of a characteristic frequency or photon energy, may lead to either an emission line or an absorption line. Which results depends on how temperature varies in the radiating surface layers—surely a fine detail—and not on the physical processes that produce and absorb radiation. The spectra of stars have many lines of each kind.

These spectral lines appeared to have clear implications. First, their only possible origin was in magnetic fields whose strengths were in a fairly narrow range centered around 3×10^{12} gauss. Such fields are found only on neutron stars. This was direct evidence for the origin of gamma-ray bursts on magnetic neutron stars. The narrow range was somewhat surprising, because common sense, and experience with pulsars, might suggest that the magnetic fields of a collection of neutron stars should be scattered over a broad range, but astronomy contains many such surprises, not understood but undoubtably true.

The second implication was that the matter that produced the gamma rays (more precisely, the spectral lines in the gamma-ray spectrum) was not moving at speeds close to the speed of light. If it were moving that fast, the energy of the cyclotron-line photons would be Doppler-shifted to energies many times higher if the radiating matter were moving toward us, and to energies many times lower if it were moving away. In a broad inflow or outflow, such as a wind

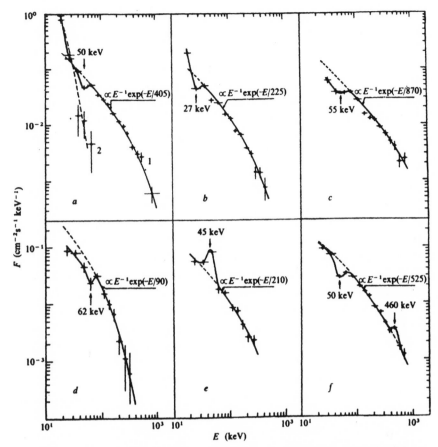

Figure 7-2. Some of Mazets's data showing apparent cyclotron lines (and, in *f*, also an apparent annihilation line) in the spectra of gamma-ray bursts. The data are shown by error bars; solid and dashed curves are fitted mathematical expressions. (Reprinted by permission from *Nature* V. 290, p. 380 © 1981 Macmillan Magazines Ltd.)

from a star or an explosion on its surface, matter moving with a wide range of speeds and directions would be observed. Averaging over all this matter would spread the absorption (or emission) line over a wide swath of photon energies, inconsistent with the narrow lines reported.

Astrophysicists asked themselves what would happen if an astronomical object "tried" to exceed the compactness limit. Of course, inanimate objects have no conscious intent, and cannot "try" to do anything. This was only a colorful way of asking a sensible question, doing what physicists call a "thought experiment": What would happen if the power of a compact source of gamma rays were steadily increased, until it exceeded the compactness limit? The gamma rays would produce a dense gas of electron–positron pairs. This pair gas would, in turn, create new gamma rays as the electrons and positrons met and annihilated. If the density of this soup of particles and gamma rays were high enough (as it was in the early universe), it would approach a state that physicists call thermal equilibrium. Matter in thermal equilibrium radiates a characteristic spectrum, called the Planck (or black-body) spectrum, named for the physicist who, in 1900, first understood it. The observed spectra of gamma-ray bursts are very unlike the Planck spectrum.

In addition, an electron–positron–gamma-ray soup will not stand still. The gamma rays will move out at the speed of light, and the particles nearly that fast. If not stopped, then the Doppler shifts of the moving electrons and positrons will be so large that a narrow cyclotron line, even if originally present, would be broadened beyond recognition. The reported narrow lines excluded this possibility. The soup's rapid expansion can only be contained by a magnetic field, and the strength of field required depends on the intensity of the radiation in the gamma-ray-burst source itself, which in turn depends on how distant the burst is assumed to be.

If the sources of gamma-ray bursts are within our galaxy, then the magnetic fields required to confine the electron–positron–gamma-ray soup are less than those typical of neutron stars. The fields of about 3×10^{12} gauss inferred from the cyclotron spectral lines would be plenty strong enough.

However, if the bursts were at cosmological distances, the required fields would be hundreds of times greater. Even if such fields exist in nature (there is some evidence that they do), they would be inconsistent with the much smaller fields required to explain the cyclotron lines. This was a case of "damned if you do, and damned if you don't": if the fields were smaller, the cyclotron line might be at the right energy, but it could not be observed because the field would not be strong enough to prevent expansion at nearly the speed of light, while if the fields were large enough to prevent expansion there would be no cyclotron line in the first place. For this reason, the reported cyclotron lines seemed to imply galactic distances for gamma-ray bursts.

Detecting spectral lines in gamma-ray-burst spectra is a dicey matter. The essential difficulty is the low-energy resolution of the scintillation counters used in nearly all gamma-ray-burst detectors (Chapter 2). Several steps intervene between the absorbed gamma ray and the electronic pulse that is finally measured and recorded. Some of the incident gamma ray's energy appears as a secondary gamma ray produced within the detector, which may escape entirely. The rest of its energy is carried by a fast electron. This electron's energy must be converted to visible light. The efficiency of conversion, the fraction of the light that is collected and that falls onto the sensitive electrode of a photo multiplier tube, and the efficiency with which this light makes an electronic pulse all differ from one absorbed gamma ray to another, partly depending on where in the scintillator the gamma ray is absorbed and partly as a result of pure statistical randomness. All of these uncertain and variable factors are multiplied together to determine the size of the final electronic pulse, from which the astronomer must infer the energy of the original gamma ray.

As a result, gamma rays of the same energy may produce pulses of rather different strengths, and pulses of the same strength may be produced by gamma rays of rather different energies. The result may be compared to a snowfall blanketing a landscape. Objects—rocks, logs, steps, even parked cars (after a real blizzard)—have their outlines blurred, just as a narrow gamma-ray spectral line will turn into a broad hump in the measured spectrum.

The scientist looking for gamma-ray spectral lines faces the same obstacles as someone trying to figure out what is under the snow from the shape of the snowy surface. Doing this is called deconvolution. It is always treacherous. Sometimes, it becomes what mathematicians call an "ill-posed problem," one without a unique solution. There are rigorous mathematical results giving optimal methods of deconvolution if the properties of the instrument and the nature of the statistical errors are precisely known. Unfortunately, real experiments are rarely so well understood. In practice, deconvolution remains, at best, a gamble. It may fail entirely. Worse, it may give seemingly plausible results that are, in fact, wrong.

The problem is easier if you have, in advance, a good idea of what form the spectrum might have, or what kinds of objects might be hidden under the snow. For example, if you know that you left a soccer ball on a smooth lawn before the snowfall, it is easy to figure out where the ball is—under the only bump. But it may be impossible to tell a round soccer ball from an oblong boulder of roughly the same size because the snow smooths the outline of the boulder. You must kick them to find out, and it may hurt.

If you know that the only thing on the lawn is a soccer ball, then the procedure by which you find it is called fitting. You compute what the snowy lawn would look like with the ball in each possible place (computers do this well), and find the location of the ball for which the computed appearance of the snowy lawn best fits what you actually see. This is not an ill-posed problem, and fitting is comparatively trouble free.

Unfortunately, the computer will always tell you the position of the ball that best fits the appearance of the snowy lawn. That is fine if there really is a ball; unless the snow is very deep it will correctly tell you where it is. However, if there is no ball, but you try the fitting procedure anyway, the computer will make its best estimate of the ball's position, telling you there is a ball where there is really only a clump of leaves under the snow, or a slight irregularity in the lawn. In the worst cases the snowfall has the same effect as a blanket thrown on a bed, which has folds, hills, and valleys, even though the mattress underneath is flat.

If you (and the computer) are cleverer, you will ask it not only where the ball is, but how big it is. If the snow is not too deep, this will work well. The computer will not only locate an actual ball, but will estimate its size accurately. If there is no ball, it will probably tell you there is one, because that hypothesis helps it fit small irregularities in the lawn or in the snowfall, but it will also tell you that the spurious ball is impossibly small, so you will ignore it.

Unfortunately, the deeper the snow, the harder the statistical problem, and the computer may find that the best fit to your lawn includes a full-sized soccer ball under the snow, even if there really is none. In fact, giving the computer the option of telling you there is a ball almost always lets it fit the surface of the snow better, so that it will almost always claim to find one. If you let it find an "anti-ball" (a hole that lowers the surface of the snow), it may find that instead.

Applied to gamma-ray-burst spectra, this means that a statistical analysis that attempts to fit to the data a theoretical spectrum that includes a spectral line will almost always find a line of some strength, either in absorption or emission (if a physical process emits photons of a certain energy, then it also absorbs at exactly that energy). With a few free parameters (numbers whose values are determined by the fitting program, such as the strength and location of a spectral line) it is possible to make almost any data fit almost any theoretical model. This is an exaggeration, but a surprisingly slight one. Such a fit does not demonstrate the correctness of the model, only the power of the fitting procedure. The famous physicist (and builder of the first nuclear reactor) Enrico Fermi once said that with three free parameters he could fit a graph of data that looks like an elephant. With five free parameters he could make its tail wag.

The scientist's task is to decide whether the fitted parameters represent physical reality (a real soccer ball under the snow, or a real spectral line in the radiation from a gamma-ray burst) or are only artifacts of the fitting procedure. Unfortunately, there is no general method of assessing this. These questions are statistical, and statistics never provide absolute answers, only likelihoods or degrees of confidence that depend on uncertain assumptions. In the case of gamma-ray-burst spectra the problem is difficult because the energy resolution of the detectors is poor (equivalent to a thick blanket of snow). Deconvolution procedures often were not clearly documented, making it impossible to assess their reliability.

No more recent experiment has confirmed the gamma-ray spectral lines reported by the Soviet and Japanese experiments. None has been demonstrably inconsistent with them either. The reality of the lines has been accepted by those who fit them into a consistent theoretical model of gamma-ray bursts (necessarily, at galactic distances), and doubted by others (those who believe bursts are at cosmological distances). The criteria used have been consistency with other evidence, or perhaps just prejudice, but not proof or disproof of their statistical significance or reexamination of the methods used to analyze the data.

8

False Light

Astronomers like to observe in visible light. Most of the major astronomical discoveries of the last half century have been made in other parts of the electromagnetic spectrum: quasars, neutron stars, and the cosmic background radiation in radio waves, black holes in X rays, and gamma-ray bursts (of course) in gamma rays, but astronomers' immediate response has been to try to observe them in visible light. This is not just nostalgia for old-fashioned methods. Observations in visible light are powerful tools because they contain much more detailed information than those at other wavelengths. Visible-light spectra may contain thousands of lines, disclosing whether the object observed is at cosmological distances, if it is rotating, if it has a magnetic field (and how large that field is), whether it orbits a companion in a binary stellar system (if so, its speed and the size of its orbit may be measured, from which the masses of the stars may be calculated), whether it is ejecting a wind or accreting infalling matter, how hot it is, and what it is made of. It is possible to point an optical telescope at a target and observe its variations for hours, night after night. Variations in its brightness tell if a star is pulsating, being eclipsed by a companion, subject to internal disturbances or eruptions, or covered with bright or dark spots. These details often cannot be found in any other way; radio, X-ray, and gamma-ray observations whet the astronomer's appetite, but optical observations are his meat.

Three difficulties delay his dinner. The first is that there are an enormous number of faint stars and star-like points of light on the sky. These all look the same, except for brightness, in a photograph (or its modern electronic equivalent). Radio, X-ray, and gamma-ray observations generally produce only approximate positions, insufficiently accurate to decide which of these faint points of light justifies detailed investigation. There are not enough telescopes to study dozens or hundreds of possible candidates individually, so it is necessary to have an accurate enough position that the optical counterpart be obvious. The second difficulty is that an astronomical object that is luminous in some other part of the electromagnetic spectrum may emit too little visible light to make it stand out on the sky, or even to observe at all. If it is comparatively bright, it may be the brightest point of light whose position is consistent with the radio, X-ray, or gamma-ray data, and the obvious target to study, but if it is faint, there is nothing to distinguish it from an enormous number of faint stars of no interest. Finally, a transient object like a gamma-ray burst may emit visible light only around the time it emits gamma rays. It may be gone long before the astronomer points his telescope in its direction. It is necessary to look not only in the right direction, but at the right time.

An optical telescope is about the worst possible instrument for observing an unpredictable event like a gamma-ray burst, because it views a tiny fraction of the sky at any time. This is sometimes described as looking through a soda straw. Worse, photographic emulsions record only the total light accumulated at each point on the sky during an exposure, typically many minutes or even hours long, and do not distinguish between a steady star and a brief flash. More modern electronic devices can produce a series of shorter exposures, but only if they are deliberately made to do so by reading out their accumulated data frequently. This increases the noise, so this is done only when the objective is specifically the observation of brief transient events. Finally, the sky is full of stars, and a visible gamma-ray burst would not look different from an ordinary star on a single exposure.

In 1978–80 an interplanetary network of detectors determined the positions of several gamma-ray bursts reasonably well, in a few cases to better than an arc-minute. The best of these positions were accurate enough that there would be very few ordinary stars detectable within the small patch of sky, known as the error box (because it contains all the possible positions allowed by the likely errors of measurement) from which the gamma-ray burst could have come. Optical astronomers set to work looking for visible counterparts to the bursts. It was easy to turn modern powerful telescopes in these directions, but nothing of interest was found. There were a few faint stars, but these apparently had nothing to do with the bursts other than accidentally being in the error boxes. Unfortunately, it took months to determine the gamma-ray-burst coordinates from the interplanetary network data. These observations constrained the brightness of persistent counterparts of the bursts months later, but said nothing about what the source of a gamma-ray burst might look like in visible light during the burst itself or shortly thereafter.

Fortunately, astronomers had faced many of these problems long before in the study of variable stars. These generally do not disappear entirely, but many of them do vary unpredictably. Tens of thousands of variable stars have been catalogued. There are only enough telescopes available to study a few of them individually at any time. In addition, when a new and particularly interesting variable star is discovered, astronomers would like to know how it varied before its discovery, a seemingly impossible task, as we cannot go backward in time.

Both objectives are met by a systematic program of photographing the entire sky, piece by piece, and doing it over and over. The resulting enormous library of photographs is saved in a permanent archive. Each position on the sky is observed repeatedly, perhaps every few months. The photographs contain data on every star in the sky, including variables not yet known to vary! If an interesting object, such as a new variable star or a gamma-ray burst, is later discovered, it is possible to go to the archives and see how bright it was and how its brightness varied before its discovery.

Astronomers realized the importance of archives of sky photographs in the late 1880s, when chemists developed photographic emulsions sensitive enough to make astronomical photography practical. Several observatories established programs of regular photography of the entire sky to create these archives. The most extensive collection is at the Harvard College Observatory, and extends from 1889 to the present. There is a gap from the early 1950s to 1970 when a director not interested in variable stars suspended the program (this still arouses strong feelings among variable-star astronomers).

Observations from a regular program of observing the entire sky are sufficient to answer several important questions. Even a single pre-discovery observation (as astronomers describe archival data from times before the object was recognized as interesting and worthy of study) with a brightness different from the post-discovery brightness might establish that the object changed in some fundamental and irreversible manner. If a few dozen observations were available, it would be possible to say if it was steady in brightness, or varied, and, if it varied, to characterize the variations as periodic (regular), consisting of occasional eruptions from a steady base, completely irregular, or in some other category. Most newly discovered variable stars turn out to belong to well-studied varieties. Establishing which variety tells a great deal about the new discovery. Further, even though the star's variations may be very slow, perhaps over decades, in a century of archives these variations are immediately available for analysis, rather than requiring additional decades of observations to accumulate.

There is a price to be paid for photographing the entire sky. Because of the laws of optics, large telescopes take photographs of small pieces of sky, and small telescopes take photographs of large pieces. Old telescopes were particularly limited because they were required to have long focal lengths (distances from the lens to the focus) to reduce aberrations (improper focusing), a problem nearly eliminated by modern optical design. To survey the entire sky in a reasonable time, or reasonably often, one must use a small telescope, and very faint stars will not be observable. It is also possible to increase the survey rate of a larger telescope by making the exposures short, but this reduces their sensitivity and increases the cost in labor and materials. In the Harvard archives exposures were typically about an hour and the telescopes ranged in size from ½ to 24 inches in diameter. Most of the useful data were obtained with very small telescopes of ½ and 3 inches in diameter. Even if a large telescope took the same number of photographs, its field of view would be so small that only a very few of its photographs would include the star, gamma-ray burst, or other object of interest.

Such tiny telescopes are closer to the size of children's toys than even to serious amateur telescopes (which typically range from 4 to 12 inches in diameter), but they can still obtain useful data. They were placed at dark sites, far away from city lights. Long exposures, typically about an hour, enabled them to detect stars as faint as 14th, 15th, or 16th magnitude on the astronomers' brightness scale. On this scale larger numbers mean dimmer objects, with the brightest star (other than the Sun) about magnitude -1, the faintest detectable with the naked eye of magnitude 6, and the faintest observable by the Hubble Space Telescope around magnitude 27. The magnitude scale is logarithmic, meaning that an increase of 1 magnitude corresponds to dividing the brightness by a factor of 2.512, an increase of 2 magnitudes to dividing it by a factor of $2.512 \times 2.512 = 6.310$, an increase of 5 magnitudes to dividing it by a factor of $2.512 \times 2.512 \times 2.512 \times 2.512 \times 2.512 = 100$ (exactly; that is where the 2.512 comes from), and so forth. To modern professional astronomers, 15th magnitude is fairly bright.

Bradley Schaefer, then a graduate student at MIT, realized that the archives of sky photographs might be useful in the study of gamma-ray bursts, which are, of course, a (very special) kind of variable star. Suppose gamma-ray bursts were repetitive events, which occur many times in the life of whatever object makes them. Then visible light from a previous gamma-ray burst might have been recorded on an archival photograph and might be identifiable because it would be at the same position on the sky as a more recent burst whose position was determined from gamma-ray data. This would be almost as good as actually catching one in action. If no such events could be found, it would at least be possible to set, statistically, an upper bound on how frequently the sources of the gamma-ray bursts become bright enough to be detected in the archival photographs.

Schaefer proceeded to the laborious task of examining thousands of archival photographs for any image at the positions of three well-localized gamma-ray bursts. To everyone's surprise, in 1981 he reported that on a photograph taken November 17, 1928, there appeared a star-like object at the position of a bright gamma-ray burst of November 19, 1978 (Figure 8-1). In fact, on that night in 1928 six identical 45-minute exposures were taken in succession of that region of sky. Only the fourth showed the new star-like object, implying that it was a brief transient, appearing and disappearing within the 45 minutes of a single exposure. The new object appeared in the image as bright as a star of 10th magnitude, but stars shine steadily, throughout the 45 minutes of exposure. If the transient's duration had been 10 seconds, typical of a gamma-ray burst, it would have been bright enough to see with the naked eye, had an eye been looking in the right place at the right time (extremely unlikely, given the rarity and unpredictability of bursts).

The actual duration of the visible flash could not be measured quantitatively, but it was possible to set a rough bound. Clearly, it was shorter than the duration of the exposure, but it was possible to make a stronger statement than that. Telescopes move to compensate for the motion of the stars across the sky (resulting from the rotation of Earth), but this compensation was not accurate on the long-ago small telescopes used to survey the sky, so the stellar images were visibly smeared. However, the image of the transient was not smeared in this manner, implying that its duration was only a small fraction of the 45-minute exposure. This was just what one would expect of a gamma-ray burst, whose duration in visible light might be about the same as its duration in gamma rays, in the range from less than a second to a few minutes.

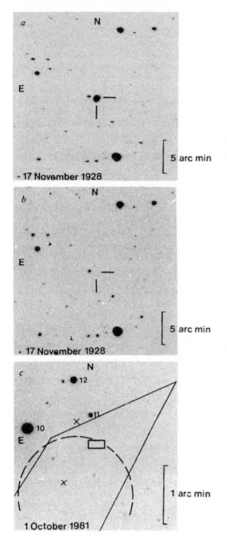

Figure 8-1. Top and middle are archival photographs from 1928. Top shows transient with two short tick marks pointing to it; middle, taken 45 minutes earlier, does not. Note that the transient's image has a different shape than the images of stars. Bottom is a modern photograph, on a much larger scale, taken by a large telescope. The small rectangle is the error box of the 1928 transient. It is inside the solid lines that show the interplanetary network error box of the gamma-ray burst of November 19, 1978. The bright star 10 is plainly visible near the transient on the archival photographs and the fainter star 12 barely so. The circle and crosses show an X-ray error box and radio positions now believed to be unrelated objects. Astronomers usually use photographic negatives in which the sky is white and stars are black because some information is lost in making a positive print. (Reprinted by permission from *Nature* V. 294, p. 723 © 1981 Macmillan Magazines Ltd.)

Schaefer then examined the sky around four more well-localized gamma-ray bursts, and in 1984 reported finding two more visible counterparts on the archival photographs. A transient recorded in 1901 corresponded to the position of a burst on November 5, 1979, and a transient recorded in 1944 corresponded to the position of a burst on January 13, 1979. Their properties resembled those of the first transient.

There were a number of implications of these results. The first was simply that if gamma-ray-burst sources produced visible flashes in the past, they should do so in the future. More modern instruments, either surveying the entire sky or concentrating on the positions of known gamma-ray bursts, should be able to catch one in the act (not just retrospectively) and study it in detail. This is remarkably difficult to do. Continuously monitoring a single small patch of sky, such as the position of a past optical flash or gamma-ray burst in the hope that it might repeat, requires a dozen or more telescopes, distributed around the world, because at any time most of them are in daylight, twilight, moonlight, under clouds, on the wrong side of Earth, or simply malfunctioning. The search for visible counterparts to the bursts themselves became the holy grail of gamma-ray-burst astronomy, but was not successful until 1999.

A second implication was that a single source produced gamma-ray bursts repetitively. In studying the positions of seven bursts, and reporting three counterparts, Schaefer examined images totaling 2.7 years of exposure to the sky. Simply dividing these numbers shows that the mean repetition rate of flashes must be about one per year (1.1 per year with a large statistical uncertainty). The reported flashes, on average, occurred about a half-century before the gamma-ray bursts, so that the period of activity must be at least this long. The sources must therefore undergo, roughly, at least fifty flashes in their lifetimes, and perhaps many more. Whatever was making the flashes (and, likely, the gamma-ray burst itself, although there was no direct evidence a gamma-ray burst occurred with each flash) could not be very energetic or destructive. This pointed strongly toward models in which gamma-ray bursts were comparatively low-energy and nearby events.

A third implication supported this inference. By comparing the brightness of the visible flashes recorded photographically to the energy measured in the actual gamma-ray bursts, it was possible to compute the ratios of gamma-ray to visible-light energies. To interpret these ratios required assuming that the events that produced the archival flashes were similar to the actual gamma-ray bursts observed by satellites decades later. There was no direct evidence for this, but none against it, either, and it was a natural assumption to make.

The ratios of gamma ray to visible energy were then found to be about 1000 for each of Schaefer's three flashes. This fitted well into the interpretation of gamma-ray bursts as something that happens on nearby neutron stars in our galaxy. A substantial fraction of neutron stars (and more than half of all stars) are binary, meaning that they consist of two stars orbiting each other. A few of these binaries are so widely separated that they can be seen to be double through a telescope. Others are so close together (or so distant from us) that they merge into a single point of light, but their binary nature may be deduced from their spectrum, which contains the lines of two stars of different types, or which shows the orbital motions of the stars around each other.

When a neutron star, black hole (or any star) in a binary emits energy, some of that energy falls on its companion star, heating it, just as the Sun illuminates and warms the daytime hemispheres of the planets. As a result, one hemisphere of the companion is hotter and brighter than the other. If it (like a planet) produces little light of its own, the difference may be dramatic. This is particularly likely if one member of the binary is a neutron star or black hole, because their strong gravity makes them very luminous if even a small amount of matter falls onto them. In extreme cases, such as the X-ray source Hercules X-1, the companion star has a hot side (facing the neutron star) and a cooler side (facing away) differing by thousands of degrees in temperature and several fold in brightness. We are used to the fact that it is warmer in the daytime than at night, but Hercules X-1 takes this to extremes! If the two stars are close together, then typically about one-thousandth of the neutron star's or black hole's luminosity, mostly emitted as X rays, appears as visible light reradiated by the illuminated hemisphere of its binary companion.

If gamma-ray bursts are produced by neutron stars, it is likely that some of them have binary companions. Some of the energy of the bursts would be reradiated by the companions. This was a natural interpretation of Schaefer's observations. The estimated ratios of gamma ray to visible intensity were about right, and certainly consistent with this idea, allowing for several uncertainties: the separation and sizes of the stars, the assumptions that the hypothetical burst energy at the time of the archival flash was the same as that of the more recent burst actually observed, the distance to the source (which determines how much of the reradiated energy appears as visible light rather than ultraviolet or infrared), and so forth. The case for a local origin of gamma-ray bursts was looking strong.

However, the inferred repetition rate of roughly one flash per year, per gamma-ray-burst source, led to trouble. This would require a great deal of effort to verify with optical telescopes, but here gamma-ray-burst detectors have an advantage. They do not suffer from daylight, moonlight, or clouds. More important, because of their lack of angular discrimination they respond to gamma rays from the entire sky (except the fraction, never as much as half, blocked by Earth) at all times. If the gamma-ray bursts themselves, and not just the visible flashes, repeat roughly once per year, then these repetitions should be easily detectable. As long as only a few bursts had been observed, with poorly determined locations, this could not be tested, but through the 1980s more and more burst positions were measured. No repetitions were found (or have ever been, even today).

This grew more and more embarrassing. The absence of repetitions among the gamma-ray bursts was possible to explain if the actual bursts were much less frequent than the optical flashes. This might be so if, for example, the gamma rays were usually beamed so that none of them were emitted in our direction, but enough fell on the companion star to heat and brighten it. Such an explanation would certainly be contrived, and perhaps implausible. If correct, it broke the connection between the gamma-ray and visible-light energies, which had seemed to fit so well with the binary neutron star model. Without some such explanation, the absence of repetitions in the gamma-ray data directly impugned the connection of the flashes to gamma-ray bursts. They might be a different phenomenon produced by the same sources, produced by different objects entirely, or possibly completely spurious. Each of these possibilities would mean that it would not be feasible to learn anything at all about gamma-ray bursts by studying the flashes.

There was a persistent skepticism toward the optical flash data themselves. Photographic emulsions contain occasional defects, some of which may resemble stars. Schaefer had also examined archival photographs of regions of the sky, many times larger, in which no gamma-ray bursts had ever been reported, and had found no flashes. This tended to dispel doubts that his transients were only defects in the emulsions. Defects might be common, but there was no reason for them to be found preferentially in images of portions of the sky from which gamma-ray bursts would be observed decades later.

There is a natural tendency among scientists to assume that newer is better, and therefore that old data are suspect. Knowledge of how old data were obtained, and their possible errors (never completely recorded in the published literature, because it would be impossible to anticipate all concerns that might develop in the future), is forgotten. We are taught that the test of a scientific result is whether it can be replicated, but we cannot go back to 1928 to repeat an observation made then. The best we can do is to perform an analogous series of observations today and compare the results statistically. Efforts were made, but did not come close to repeating the near-century of sky monitoring found in the archives, and no useful conclusions could be drawn.

Anna Żytkow is an American astrophysicist of Polish origin who has worked at several institutions in the United States, Poland, and Britain. She made a strong impression on a number of her male colleagues in that woman-starved field, one of whom described her ("brilliant and beautiful, with long black hair . . .") decades later on the World Wide Web. Interrupted by a nearly fatal mountaineering accident from which recovery took many years, her career was not very successful, but she did play an important and controversial part in the debate over the archival optical flashes. Several years after Schaefer's work she reexamined the photographs in painstaking detail. Finding some anomalies in the images he interpreted as flashes, she concluded that "we should treat with great caution the suggestion that gamma-ray bursts are necessarily accompanied by optical flashes." That is a polite way of saying "Don't believe it." Schaefer then also closely reexamined the images, but he replied that "fundamental errors of methodology and data analysis are identified which invalidate all the major points raised by Żytkow."

The bitterness of this quarrel is shown by the fact that 19 months elapsed between the submission of Żytkow's paper, finally published in 1990, and its acceptance (2 to 4 months is typical). The journal editor sent Żytkow's paper to Schaefer for comment. Schaefer then wrote his own paper in reply, which was sent to Żytkow. Everything was then sent to a neutral referee whose identity was (and still is) secret. There were several rounds of argument. The papers were eventually published, but the disagreement was never resolved. Perhaps the best summary was that of the neutral referee assigned the onerous task of assessing Schaefer's and Żytkow's manuscripts and revisions and lengthy discussions of each other's work: "archival plates* were never intended for the detection and analysis of fast optical transients. For this reason they have severe, fundamental limitations. . . . This simple fact has either been forgotten or ignored. . . . Most distressing is the descent into enmity."

Such inconclusive results are remarkably common in science. Often a controversial result is neither directly confirmed nor disproved. Occasionally, as for the archival flashes at the positions of gamma-ray bursts, this is because it involves a past natural event that cannot be reproduced at will. Related examples include ball lightning and light and sounds and peculiar animal behavior reported to occur before or during earthquakes. Some, or all, of these may be fictitious, but sim-

---

*Astronomical photographs are made on glass plates rather than film because film stretches, precluding precise measurements of the positions of stars.

ilar skepticism was applied to meteorites until about 1800 and they are certainly real! More often, disputes remain unresolved because scientists' interests change, and the old controversies become less pressing and no longer justify the effort required to settle them. We will probably never know, for sure, the origins of the reported flashes from 1901, 1928, and 1944, but our more modern understanding makes it unlikely that they were directly associated with the gamma-ray bursts of 1978 and 1979 at whose positions they were found. However, that is getting ahead of the story by a decade.

For most gamma-ray-burst astronomers the issue ended with the standoff between Schaefer and Żytkow in 1990. But there was more. Further attempts to find archival counterparts were unsuccessful, and another reexamination of Schaefer's images (by the German astrophysicist Jochen Greiner in 1992) was also skeptical of their reality. A group of astronomers led by René Hudec in the Czech Republic kept looking, and found, on a photograph taken in 1905, an apparent counterpart to a gamma-ray burst that occurred in 1991. In 1994 they published a very detailed examination of the image. This did not receive much attention, partly because the earlier controversy had left the whole subject of archival counterparts under a cloud, and partly because by the mid-1990s new gamma-ray data had answered the question of the distance scale. Some years later, analysis of additional data led to a refinement of the measured position of the 1991 gamma-ray burst, and it was shown that the supposed 1905 transient was not in the same place on the sky. This result was not even published. Archival counterparts no longer seemed so important.

If, in the year 2000, you had asked a gamma-ray-burst astronomer what he thought of the visible light flashes reported on archival photographs at the positions of gamma-ray bursts, he would almost certainly have said either that he did not believe that they were real, or that, even if real, they have nothing to do with gamma-ray bursts. Yet they have never been conclusively shown to be spurious, or directly disproved by an independent equivalent repetition of the observations. This seems to contradict the customary picture of how scientific disputes are resolved: repeat the experiment.

Occasionally that is possible, and is done with conclusive results. Direct replication is most often used in fields, particularly in the biological sciences, in which experiments are comparatively quick and cheap and there is unlikely to be a well-established theory relating the results of one experiment to those of others. More often, especially in the physical sciences, an indirect argument is used implicitly: if the controversial result has consequences that are contradicted by other well-established results, then it is deemed unreliable and ignored. Consistency may substitute for replication because each tests the controversial result, consistency indirectly and replication directly.

Both archival flashes and lines in the gamma-ray spectra of bursts were tested by consistency checks rather than by replication of the experiments. Waiting for another century of sky monitoring was not feasible. A better gamma-ray spectroscope could have been built and flown, but was never in the budget. In each case it took a decade for the inconsistency to become apparent, and through that decade each was considered strong evidence for the local origin of gamma-ray bursts.

It is familiar in science that when an experiment is in error the source of the error is not found. Were it easy to find, the original experimenter would have caught it himself. More often, it is buried deep within a complex experiment, in some unanticipated manner, perhaps as a subtly flawed assumption or component, hidden even from close inspection. The difficulty of finding the source of error, even when we are confident there has been an error, is the reason replication is considered the ultimate test. Usually, though, we know that there was an error only because the results are inconsistent with a greater body of independent and mutually consistent experiments.

To be trustworthy a scientific result must be consistent with other well-established results. Logical consistency is the most fundamental law of nature assumed by science. Where does this leave a genuinely new discovery? Usually, a new discovery is completely consistent with previous empirical results; it simply shows that a new experiment, never previously performed, found a new phenomenon. Only the theoretical inferences drawn from it overthrow previous theoretical ideas; the results of previous experiments are not contradicted. For example, the theory of relativity displaced Newtonian mechanics. The experiments from which Newtonian mechanics was obtained were correct and remained valid; the effects of relativity could only be seen in newer experiments involving very rapidly moving particles.

9

The Copernican Dilemma

In school we are taught that Mikołay Kopernik, the Polish astronomer also known by his Latin name Nicolaus Copernicus, established that Earth and the planets revolve around the Sun, rather than the planets and Sun revolving around Earth, as had generally been assumed since the time of the ancient Greeks. The importance of Copernicus's ideas was both philosophical and scientific: Man is not at the center of the universe, but is only an insignificant spectator, viewing its fireworks from somewhere in the bleachers.

In modern times this has been elevated into the "cosmological principle," which states that, if *averaged* over a sufficiently large region, the properties of the universe are the same everywhere; our neighborhood is completely ordinary and unremarkable. We are not special, and our home is not special, either. This is one of the foundations of nearly all modern cosmologies. Originally almost as much a matter of faith as the ancient doctrine that Earth is at the center of the universe, it is now supported by studies of the cosmic microwave background radiation, a relic of the Big Bang, which is remarkably uniform.

The cosmological principle only applies to average properties, and only to averages taken over sufficiently large volumes. For example, it is clear that the mean density of Earth, 5.5 grams per cubic centimeter, is much greater than the mean density of the universe, which is roughly 10^{-28} grams per cubic centimeter. However, as the average is taken over larger and larger volumes, the contribution of dense bodies near us becomes less and less important. Eventually the computed average density approaches the mean density of the universe, although just how large a volume must be averaged over remains controversial; studies of the distribution of galaxies in space suggest that the averaging must be carried out over at least 10^8 light-years, and perhaps more.

These esoteric questions are closely related to the problem of determining the origin and distances of gamma-ray bursts. For the first quarter-century after their discovery (until 1997) their distances were not measured directly, because no gamma-ray burst had ever been convincingly and correctly identified with another astronomical object whose distance might be known or determinable. As we have seen in Chapter 8, attempts to identify bursts with brief flashes of light in our galaxy turned out to be misleading. This left astronomers with only indirect and statistical methods.

The first statistical method relied on the measurement of the distribution of the locations of gamma-ray bursts on the sky. Although these locations were not determined (with the exception of the March 5, 1979, event) accurately enough to point to individual identifications, even crude positional data could be useful statistically. For example, a concentration of locations in the galactic equator (the midplane of the galactic pancake) would convincingly establish that the bursts come from within our galaxy. A single location, however accurate, could not establish this, but a hundred locations scattered within 30 degrees of the galactic equator would, even if the uncertainties in the individual positions were very large.

The actual data gave no help at all. It was recognized as early as 1974, in a paper published by the Los Alamos group in a follow-up to their discovery of gamma-ray bursts as a new astronomical phenomenon, that they show "no tendency to cluster around the Galactic center, nor even along the Galactic equator." Further data only refined this conclusion. For example, Figure 9-1, published in 1987 (but based on data collected in 1978–80 by a French–American–Soviet interplanetary network) clearly indicates that all directions on the sky are equally likely to be the origin of gamma-ray bursts.

The eye and mind can often be deceptive, appearing to find patterns in random data (such as the several close pairs or the three points in a straight line in the upper right of Figure 9-1), but none of these patterns is anything other than the consequence of pure chance. There are no statistically significant preferred directions or deviations from uniformity in these data. Trained scientists have been led astray because true random data appear unexpectedly irregular and nonuniform; the human eye and mind are so good at finding hidden regularities and patterns, a very useful skill when they really are present, that they often find patterns where none exist. Ask a

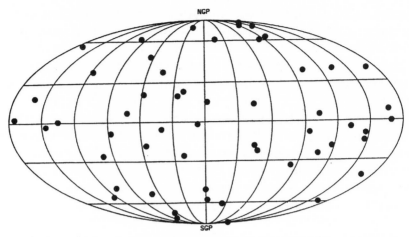

Figure 9-1. Positions of gamma-ray bursts on the sky obtained from the First Interplanetary Network 1978–80. The bursts are plotted in galactic coordinates, in which the galactic plane (Milky Way) is a horizontal band across the middle of the figure and the galactic center is at the center. This is an equal-area projection, so that if bursts are isotropically distributed on the sky they are equally likely to be anywhere inside the outermost curve of the figure. (Reprinted by permission from *Astrophys. J. Suppl.* V. 64, p. 382 © 1987 American Astronomical Society.)

gambler. To this day, after the detection of several thousand bursts, and despite earnest efforts to show the contrary, no deviation from a uniform random distribution (isotropy) in the directions of gamma-ray bursts on the sky has ever been convincingly demonstrated.

It is hardly surprising that statistical studies of the directions from which gamma-ray bursts come should tell us something about how they are distributed in space. More remarkable is the fact that studies of their brightnesses alone can yield equally important spatial information. In particular, comparing the numbers of bright and faint objects (gamma-ray bursts or anything else), without any knowledge of where they are on the sky, permits us to compare their density near us to that in more distant space.

The argument is based on simple geometry. The volume of a sphere is proportional to the cube of its radius; this scaling is also true of a hemisphere, a cube, a regular tetrahedron, an elephant, or any other three-dimensional object. If you double the radius of a sphere, you increase its volume eightfold. If the number density (how many per unit volume) of objects being studied is uniform in space, the larger sphere will contain eight times as many of them as the smaller sphere.

The brightness of any source of radiation varies inversely as the square of its distance, assuming the radiation passes through space without absorption. This is called the inverse square law. As a result, the objects in the larger sphere, with twice the radius, will, on average, be one-quarter of the brightness of those in the smaller sphere.

Combining these simple geometrical laws, we estimate the number N of identical objects that appear brighter than some minimum value S. The minimum determines a distance R at which the object's brightness just equals S and it can barely be detected; all objects closer than R can be observed, but none farther away. The number N of the objects inside this sphere of radius R is proportional to its volume, so that we write $N \propto R^3$, where the symbol $\propto$ simply means "is proportional to." The constant of proportionality depends on the actual number density of the objects, but we can ignore it. The distance R is inversely proportional to the square root of the minimum detectable brightness ($R \propto S^{-\frac{1}{2}}$) because of the inverse square law. Combining these relations, the number of objects brighter than the minimum is inversely proportional to the $\frac{3}{2}$ power of that threshold, or proportional to the $-\frac{3}{2}$ power: $N \propto S^{-\frac{3}{2}}$. A fourfold reduction in the minimum brightness corresponds to a twofold increase in R and an eightfold increase in the number of sources detected.

This result is simplest to derive if all the sources radiate the same amount of radiation, as we have assumed, so that for a given minimum S there is a single limiting distance R. The same result applies even if the sources produce different amounts of radiation, or if their radiation is beamed in random directions. It even applies to variable sources and to transient objects such as gamma-ray bursts. Any quantity that follows an inverse square law dependence on distance may be used for S, and for gamma-ray bursts S is usually the cumulative radiation received throughout the burst (called the fluence, measured in units of energy per area of absorbing or detecting surface), rather than the brightness at any given time (called the flux, or intensity, measured in units of energy per area per second).

Number-flux (N vs. S) relations were introduced into astronomy in the 1960s. At that time radio astronomy was blossoming, but the early radio telescopes were not capable of accurately measuring the positions on the sky of the newly discovered sources of radio emission. With only a few exceptions, it was not possible to associate the radio sources with astronomical objects observed in visible light. It was also not possible to estimate their distances from the radio observations directly, because their radio emission did not contain any characteristic signatures of distance, such as red shifted spectral lines.

The exceptions were the newly discovered quasars, the most powerful and distant objects in the universe (except for gamma-ray bursts, which were yet to be discovered), and astronomers were eager to learn more about the rest of the radio sources. Radio astronomers faced the same problem as gamma-ray-burst astronomers, but more than a decade earlier. The radio astronomers solved it by inventing the N vs. S test, which in turn was based on statistical methods used early in the century in the study of the distribution of stars in our galaxy, at a time when even the distances to ordinary stars were poorly known.

If our assumptions are valid, N is predicted to vary as the $-\frac{3}{2}$ power of S. Increasing the sensitivity of an astronomical instrument (reducing the minimum brightness it can detect) tenfold is predicted to increase the number of objects detected by a factor of about thirty; increasing the sensitivity a hundredfold is predicted to increase the number of objects detected by a factor of a thousand, and so forth. This certainly accords, at least qualitatively, with our experience of the night sky: there are many more faint stars than bright ones.

That is the theory. Unfortunately, the data do not confirm it. The mathematics is simple and unimpeachable, so one of the assumptions must be wrong. In the 1960s it was discovered that the number of faint radio sources observed was even greater than predicted by the $-\frac{3}{2}$ power law. The explanation was that the number density of radio sources was greater far from us than nearby.

This appears to violate the cosmological principle. It is explained by the time required for signals to reach us from very distant objects. We observe distant space as it was long ago, when the radiation we see was emitted. The excess of faint radio sources implies that they were more abundant in the distant past, when the universe and everything in it were younger. A purely statistical argument thus told us something important and unexpected about the early universe.

The excess of faint radio sources delivered one of the very few clear observational verdicts on a cosmological model. In the steady state cosmology, then popular, the universe has no beginning and no end and is always and everywhere infinitely old. Matter was supposed to be continually created from empty space as the universe expands, keeping the abundance of radio sources (and everything else) constant. The radio source counts showed that this is wrong.

It was realized early, in the same paper quoted earlier in this chapter, that these methods could be, and needed to be, applied to gamma-ray bursts to determine how they are distributed in space, for there was little prospect of direct distance measurements. Unfortunately, the initial data from the Vela satellites were too scanty and the instruments not sensitive enough to justify confident conclusions.

Within two years a team at NASA Marshall Space Flight Center in Huntsville, Alabama, led by Jerry Fishman, had designed and built instruments capable of answering this question. These instruments consisted of very large NaI(Tl) crystal scintillators, some commercially produced for use in medical gamma-ray imaging. Their large size was crucial, for it would enable them to detect gamma-ray bursts too weak to be detected by any other instrument, just as fainter stars may be seen with a big telescope than a small one. The ultimate goal of this project was an instrument to be called BATSE (Burst and Transient Source Experiment) that would be built for launch into space aboard the Gamma-Ray Observatory satellite (Chapter 11), but these preliminary versions could be flown on high altitude balloons. Although the atmosphere absorbs gamma rays, so that gamma-ray bursts cannot be observed from the ground, these balloons fly over 120,000 feet high, four times higher than a commercial jet airplane. At this altitude there is so little atmosphere (less than 1% as much as at sea level) that gamma rays from space penetrate. Balloon flight is much cheaper than space flight, and may require only a year of preparation, in contrast to the decade or more of planning, reviews, and delay typically involved in a NASA space mission. Because weak bursts were expected to be much more frequent than strong bursts, even a brief balloon flight was predicted to observe several of them.

The Huntsville group flew various versions of their experiment from 1975 to 1982. In each case the expected faint gamma-ray bursts were absent or observed much less often than predicted, based on the observations of intense gamma-ray bursts by other instruments. Figure 9-2 shows their initial results, published in 1978. The predicted $-\frac{3}{2}$ power law in the N vs. S relation was not observed. This was a substantial, and statistically very significant, discrepancy. Both the deficiency of faint bursts and the isotropy of bursts in general were confirmed by data published by Mazets's Soviet group in 1981.

There was something fundamentally wrong with the assumptions. Too few faint gamma-ray bursts implies a deficiency of very distant bursts, the opposite of the situation with radio sources. But how distant were these missing bursts? That question could not be answered until the gamma-ray-burst distance scale could be established, and it remained completely unknown, with competing hypotheses differing by a factor of at least ten million (ignoring the crazier ideas, such as gamma-ray bursts in the outer solar system, which would extend the range to a factor of a trillion).

Figure 9-2. The number-flux relation observed for gamma-ray bursts. Data are shown by "error bars," vertical lines that indicate statistical uncertainties, and by horizontal hatched bars with arrows pointing downward, which indicate upper limits. The straight line is a $-\frac{3}{2}$ power law, as predicted if bursts are uniformly distributed in space, and curved lines represent various more complex distributions. The brighter bursts follow the straight line, but there is a striking deficiency of faint bursts. (*Astrophys. J.* V. 223, p. L14 [1978].)

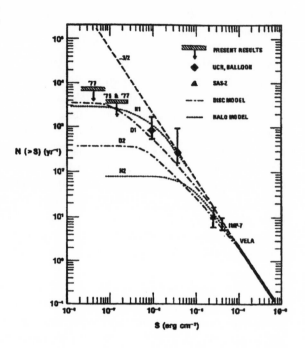

If the gamma-ray-burst sources were within our galaxy (e.g., neutron stars), then a deficiency of faint bursts would be expected if observations were sensitive enough to detect bursts at distances greater than the thickness of the galactic pancake. That is because outside the pancake there are few stars and, presumably, few burst sources, whatever they are. So, in this picture, a deficiency of faint bursts was expected. Unfortunately, this deficiency should have been observed in directions looking out of the galactic pancake, while there should have been plenty of bursts in directions looking edgewise into the pancake. In slightly more technical language, there should have been an anisotropy in the burst distribution on the sky, with an excess along the galactic equator. Figure 9-1 shows that there was no such excess.

If, on the other hand, bursts were produced billions of light-years away, from the edges of the observable universe, then the deficiency of faint bursts could be explained in analogy with the excess of faint radio sources: in far distant, and long ago, space, gamma-ray-burst sources were rarer than they are here and now. Why this should be was not understood (like the analogous, but opposite, requirement for radio sources), but for the purposes of this argument it did not need to be understood; it simply had to be possible, and it was. The cosmological principle meant that a line in any direction should pass through similar, on average, regions of space, with similar densities of gamma-ray-burst sources, whatever they were, so that in this cosmological picture the deficiency of faint bursts was entirely consistent with their isotropic distribution on the sky.

In fact, Vladimir Usov and a co-worker, Gennadi Chibisov, hypothesizing that gamma-ray bursts are at cosmological distances, had predicted just such a deficiency of faint bursts in 1975 (Figure 9-3). The resemblance between Figures 9-2 and 9-3 is almost uncanny. Why was this strong evidence ignored by most astronomers?

One possible explanation was that the interpretation of the balloon-borne experiments depended on the comparison of counts of weak bursts, made from a balloon, to counts of the much rarer strong bursts, made with satellite-borne instruments. There is nothing wrong with carrying an instrument on a balloon, but it is notorious in astronomy that calibration errors may make it difficult to reconcile observations made with different instruments unless there is a standard of comparison observed by both. No such standard existed for gamma-ray-burst counts. The balloon-borne instruments could not observe strong bursts because they are so infrequent, and balloon technology limited flights to a day or two in duration. Satellite instruments, in orbit for years, could observe the rare strong bursts, but were too small to detect weak bursts at all.

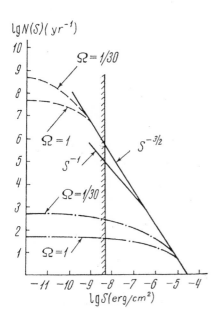

Figure 9-3. Usov and Chibisov's original 1975 prediction of the number-flux relation for gamma-ray bursts if they are at cosmological distances. Straight lines are $-\frac{3}{2}$ and -1 power laws, and the curves are predictions for various assumptions about the luminosity of gamma-ray bursts and the density of the universe (described by a parameter Ω, whose value is uncertain but is now believed to be between 0.3 and 1. "lg" is an abbreviation for logarithm (base 10). The qualitative similarity to the data of Figure 9-2 is striking. (Reprinted by permission from *Soviet Astronomy–AJ* V. 19, p. 115 © 1975 American Institute of Physics.)

It also was not clear that the directional and intensity data were directly comparable. Directions were measured for the brighter bursts, while the balloon data revealed a deficiency of fainter ones whose directions were not measured (partly because the balloon-borne instruments were not designed to measure direction, and partly because they did not find bursts at all). These problems made it possible for most scientists in the field, already leaning toward galactic neutron stars as the sources of gamma-ray bursts, to ignore or dismiss the balloon data. James Watson once said that one should not expect a theory to agree with all the data, because some of the data were likely to be wrong. His work on the double helix had briefly been thwarted by incorrect information found in a biochemistry text. Although occasionally justified, this argument is dangerous and, like habanero peppers, must be used extremely sparingly.

The Soviet data were all obtained from space and did not depend on comparing measurements made with different instruments. They may have been discounted because of a widespread prejudice in the West that while Soviet theoretical physics was as good as any in the world, Soviet experimental physics and observational astronomy were untrustworthy. This had some foundation in reality. For nearly 20 years (1976–1993) a Soviet optical telescope was the largest in the world, but because of a poorly made mirror, deficient instrumentation, and bad atmospheric conditions (for political reasons it was placed in Russia where there are no really good astronomical sites), it made no significant discoveries. Also, like Usov, Mazets was isolated in the Soviet Union and never became part of the inner circle of astronomy. A shortage of hard currency forced him to publish his crucial and lengthy 1981 papers in a comparatively obscure journal, rather than the more prestigious *Astrophysical Journal*, which charges a substantial fee (proportional to length) to publish a paper.

In addition, there was an implicit inconsistency in Mazets's data. The burst counts and distribution on the sky indicated that bursts occur at cosmological distances. But the very same instruments observed lines in their gamma-ray spectra, which could only be explained as cyclotron lines from bursts within our galaxy. It seemed likely that something was wrong, but without independent evidence it was impossible to decide rationally which data were likely to be wrong, and could be ignored, and which must be dealt with. This made it easy to dismiss or ignore the Soviet results.

The uniform distribution of burst arrival directions tells us that the distribution of gamma-ray-burst sources in space is a sphere or spherical shell, with us at the center (some other extremely contrived and implausible distributions are also possible). But Copernicus taught us that we are not in a special preferred position in the universe; Earth is not at the center of the solar system, the Sun is not at the center of the galaxy, and so forth. There is no reason to believe we are at the center of the distribution of gamma-ray bursts. If our instruments are sensitive enough to detect bursts at the edge of their spatial distribution, then they should not be isotropic on the sky, contrary to observation; if our instruments are less sensitive, then the $N \propto S^{-3/2}$ law should hold, also contrary to observation. That is the Copernican dilemma.

One possibility was that, for once, the Copernican principle was wrong. For example, suppose the galaxy is at the center of an enormous spherical cloud that contains, among other things, the mysterious sources of gamma-ray bursts. This is not a crazy idea; it has been discovered in recent decades that most galaxies contain much more invisible mass than is present invisible stars. Further, this invisible mass is mostly found outside the visible galaxy, perhaps in a giant halo. It is quite plausible that our galaxy has a similar massive halo. Although the nature of the matter in these halos is unknown, they might (why not?) contain the sources of gamma-ray bursts. If the halo is large enough (at least ten times as large as the visible part of our galaxy) and spherical enough, then it would appear to us to be the same in all directions; we would not be at its exact center, but close enough to the center to agree with the data. Experimental data are never exact. If the halo did not go on forever (it could not, because if it did it would fill the universe with more mass than it is known to contain), then the deficiency of faint gamma-ray bursts might be accounted for.

Explanations of this kind are criticized as ad hoc, meaning that they require ingredients and assumptions concocted for the occasion and without direct experimental support, although not demonstrably wrong. This is a valid criticism, but not a fatal one. The galactic halo idea remained popular for many years because it permitted gamma-ray bursts to be comparatively low-energy events occurring on magnetic neutron stars. A sufficiently large halo might even reconcile the bright March 5, 1979, event in the Large Magellanic Cloud with the fainter regular bursts, because now the regular bursts could be more distant than the cloud, rather than closer. As more data accumulated concerning the angular distribution of gamma-ray bursts, they relentlessly remained uniformly distributed on the sky, and the size and sphericity required of the assumed halo steadily increased.

The most natural explanation of the Copernican dilemma was that gamma-ray bursts were at truly cosmological distances of about ten billion light-years. As we have seen, this idea went back at least to the early days of gamma-ray-burst theory in 1975, when it was suggested by Vladimir Usov. The prejudice against cosmological distances, probably rooted in the desire to minimize the energy a model would be required to explain, was so deep that Mazets himself asked, when he first heard Usov present his work in 1974, "Is your talk a joke?" The cosmological hypothesis then ran afoul of the arguments discussed in the last several chapters for galactic neutron stars. Most astrophysicists found these arguments persuasive, although an occasional paper arguing for cosmological distances appeared in the 1980s.

There may have been other reasons why the Copernican dilemma was neglected. One was that most gamma-ray-burst observers were not astronomers by background. As a result, while reporting the distribution of bursts on the sky and in intensity, they did not dwell on the implications of these data. Fishman's 1978 paper and Mazets's 1981 paper dismissed the cosmological hypothesis. The observers left interpretation to the theoretical astrophysicists. The theorists would generally try to think of a model for the internal workings of bursts, publish it, and hope that it would later be proved right and be the tide which, taken at the flood, would lead them on to fame (if not fortune). After publishing his pet model, a theorist would go on to work on some unrelated subject. The data developed only slowly, and the theorists' short attention spans did not encourage careful consideration of all the data.

Even science is subject to fashion. Imperceptibly, and unaware, scientists come to accept the views of those around them, and to dismiss or ignore discrepant evidence, just as small children learn to speak the language, idiom, and accent they hear. In time, such a tacit consensus slowly changes, but in the early 1980s most astronomers were willing to assume that the Copernican dilemma would disappear when more data became available. Fishman's space experiment, to be known as BATSE, was planned in the late 1970s but was not launched until 1991. It was natural simply to say that it was better to wait for BATSE, which would answer all questions, than to try to interpret the limited data then available. This attitude stalled thinking about gamma-ray bursts for a decade.

The persistence of the neutron-star hypothesis was harmful, but not because it eventually turned out to be wrong. Sometimes pursuing erroneous ideas leads to discovery. This idea was harmful because the general belief that to understand bursts required understanding the complexities of neutron stars, probably undergoing magnetic flares or sucking matter from their surroundings, kept most astrophysicists from working on the simpler problems of gamma-ray bursts at cosmological distances. It might be thought that it is more difficult to explain a big explosion than a little one, but this need not be so. Dozens of different processes can explain a small release of energy, but the requirement that it be gigantic narrows the possibilities.

In addition, assuming that burst sources were only a few hundred light-years away led observers to focus on trying to determine very accurate angular positions of a few bursts, in order to find visible light counterparts. This approach worked in the 1960s in radio astronomy, where it led to the discovery of quasars, and in the 1970s in X-ray astronomy, where it led to the discovery, in our galaxy, of neutron stars and black holes with binary companions. Attempts to determine accurate positions of gamma-ray bursts consumed a great deal of effort, but were long fruitless. The Copernican dilemma was finally resolved by statistical studies of rough positions of a large number of bursts, the same kind of data that created it.

10

Soft Gamma Repeaters

The year 1986 marked the high point of the view of gamma-ray bursts as a comparatively nearby phenomenon, probably occurring on magnetic neutron stars in our galaxy. Five separate lines of evidence and argument pointed in that direction: the complexity of their light curves, the compactness argument against excessively high luminosities, the identification of the superburst of March 5, 1979, with a supernova remnant in the Large Magellanic Cloud, the reports of lines in the gamma-ray spectra of bursts, and the apparent counterparts of bursts on archival sky photographs. Only the Copernican dilemma was inconsistent with this view, and most astronomers believed it would be resolved, either by locating the bursts in an extended halo of our galaxy or by further data that would find anisotropy in the burst distribution or the missing faint bursts.

Then it all began to unravel. The first thread to come loose was the March 5, 1979, superburst. In 1986 an American–Franco–Soviet collaboration discovered repeated bursts at the same position as a burst observed January 7, 1979. Data from the early 1980s (which had not been examined in detail before) showed more than one hundred outbursts. Most were clustered in November 1983. Others were scattered, some in bunches but others singly, from 1979 through 1984 (Figure 10-1).

This was entirely unlike the behavior of any other gamma-ray-burst source, with the sole exception of the superburst of March 5, 1979, which was detected about fifteen more times in the few months following its great eruption. However, these afterbursts were so weak in comparison to its main event that they had generally been ignored, rather than being considered a characteristic distinguishing this ob-

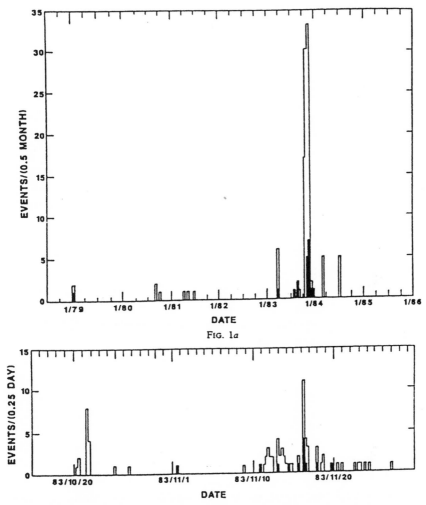

FIG. 1a

Figure 10-1. History of a soft gamma repeater, showing about one hundred outbursts. Like a volcano or comet shower, long periods of somnolence are interrupted by an occasional frenzy of activity. (*Astrophys. J. Lett.* V. 320, p. L113 [1987].)

ject from other gamma-ray-burst sources. Similar afterbursts following any other gamma-ray burst, weaker in proportion, would have been too faint to detect at all, so their absence could simply be attributed to the limited sensitivity of the instruments.

It was clear that a new type of astronomical object had been discovered or, at least, recognized for the first time. The repetitive outbursts differed from the more familiar gamma-ray bursts, now dubbed "classical," in several ways. They lasted one- or two-tenths of a second

(Figure 10-2), in contrast to classical gamma-ray bursts, whose durations ranged up to hundreds of seconds. In addition, the spectra of these new events (Figure 10-3) contained no observable gamma rays with energies above about 250 KeV, in contrast to those of classical gamma-ray bursts, which extend smoothly up to at least 100 MeV (Figure 10-4).

The new objects were soon named soft gamma repeaters. "Soft" is gamma-ray physicists' and astronomers' jargon for low energy, and describes the fact that their spectra do not contain any of the more energetic gamma rays so abundant in the spectra of classical gamma-ray bursts. "Repeater" simply says that they repeat. They could also have been called "short," for their durations, but were not, perhaps because a few classical bursts are just as short, or perhaps for brevity. On the other hand, all repeaters (with one exception) have soft spectra, and all bursts with soft spectra repeat, so it is clear that they represent a distinct class of object.

It was soon realized that the source of the superburst of March 5, 1979, was actually a soft gamma repeater: it repeated, all its outbursts were brief (except for the superburst itself), and its spectrum was soft (with the single exception of the first and most intense quarter-second of its superburst). A few other soft gamma repeaters have been detected since, all in our galaxy proper, for a total of about five (there is some uncertainty about one or two). On August 27, 1998, and again on April 18, 2001, one of them had superbursts similar to that of March 5, 1979, but roughly one-tenth as powerful.

It is possible to estimate the number of soft gamma repeaters in our galaxy that have not yet been discovered. This might seem impossible—how can you tell they are out there if you have not observed them? It can be done, reasonably reliably, if you assume that they produce outbursts with the same frequency and strength as the repeaters we do know about. Then it is possible to estimate, given how often, for how long, and with what sensitivity instruments observe the sky, the probability that a repeater would have been detected. This probability is a statistical estimate of the fraction of all the repeaters in existence that have been discovered; from this it is easy to estimate the number missed. The result for the total number of repeaters is about seven, out of about a hundred billion stars in the galaxy. Soft gamma repeaters either are very rare objects, or they are only active for a very short time (if they don't produce outbursts, they cannot be detected, and are not included in the estimate). In contrast, thousands of classical gamma-ray bursts have been observed, roughly

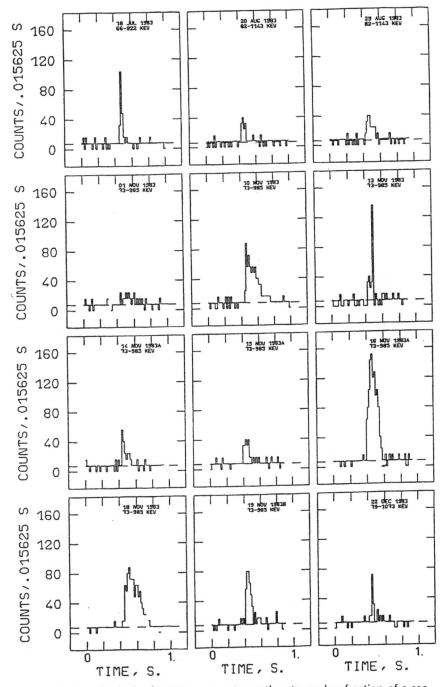

Figure 10-2. Individual soft gamma repeater outbursts, each a fraction of a second long. Compare with Figure 1-2, showing classical gamma-ray bursts. (Reprinted by permission from *Astrophys. J. Lett.* V. 320, p. L107 © 1987 American Astronomical Society.)

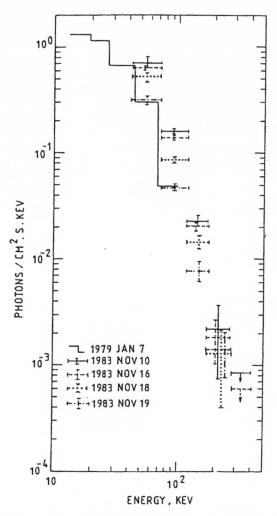

Figure 10-3. The spectrum of a soft gamma repeater. Note the complete absence of gamma rays of more than 300 KeV. (Reprinted by permission from *Astrophys. J. Lett.* V. 320, p. L108 © 1987 American Astronomical Society.)

one every day, and there is no good evidence that any of them has erupted more than once.

The importance of the discovery of soft gamma repeaters was that inferences drawn from the superburst of March 5, 1979, could not properly be applied to classical gamma-ray bursts. No longer could it be argued that gamma-ray bursts originate in or near our galaxy on the grounds that the superburst was convincingly located in the galaxy's satellite, the Large Magellanic Cloud. It could still be said that the objects now known as soft gamma repeaters are found in the galaxy and the cloud, but this was no longer evidence that classical gamma-ray bursts are galactic.

Instead, astronomers were faced with two classes of bursts to ex-

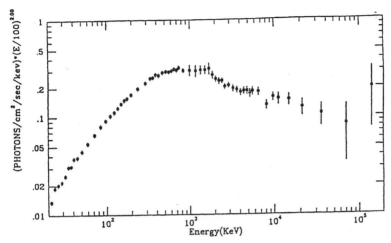

Figure 10-4. The spectrum of a classical gamma-ray burst, obtained by BATSE and instruments sensitive to higher-energy gamma rays. The data are scaled so that peaks indicate the spectral regions in which most of the energy is found. Energy is broadly distributed from a few hundred KeV up to many MeV. (*Pub. Astron. Soc. Pacific* V. 107, p. 1148 [1995].)

plain rather than one. This actually made their task easier, rather than harder, because no longer did apparently inconsistent lines of argument require reconciliation. Evidence could be divided into two classes, one pertaining to soft gamma repeaters and the other to classical gamma-ray bursts, and there was no reason to demand that conclusions about one class be consistent with those about the other.

In particular, the Copernican dilemma was much less severe. It had seemed to imply that classical gamma-ray bursts (nearly all the statistical evidence was derived from the numerous classical bursts, rather than from the handful of repeaters) were distributed over a spherical volume with a definite outer boundary, but with us at the center. This could be explained naturally if bursts were at cosmological distances, but such great distances had apparently been excluded by the identification of the March 5, 1979, superburst in the Large Magellanic Cloud. Now, the contradiction was removed because the superburst was realized to be entirely unrelated to the classical bursts.

Of the two classes of erupting gamma-ray objects, soft gamma repeaters appeared to pose the easier problem. Some of the ideas suggested in the first wave of invention after the discovery of bursts in 1973, when most astronomers assumed they were a few hundred light-years away, within the thickness of the galactic disc, could be applied to soft gamma repeaters. They were actually hundreds of times more distant, in the remote reaches of the galaxy (or its satellites), so their energy requirements were tens of thousands of times greater, but more than a billion times less than those required to explain classical bursts if they were at cosmological distances. In fact, most of these early ideas were so sketchy and uncertain that they made no specific predictions for the spectrum, duration, or time dependence of the bursts they purported to explain, and could be applied to repeaters as readily as to the classical bursts for which they were invented.

Two models seemed to be possible explanations of the soft gamma repeaters. One was based on the 1973 model of bursts as the result of comets falling onto neutron stars. Harwit and Salpeter had assumed neutron stars simply because of their intense gravity. The amount of energy released would be about a hundred thousand times greater than if the same object fell onto an ordinary star, and a billion times greater than if it fell onto Earth. A black hole might serve as well as a neutron star, if only the comet fell in off-center, so that it was not simply swallowed up without a burp (matter accreted off-center is much harder to digest because it goes into orbit, rather than smoothly going down the hole).

A large comet falling onto a neutron star might explain a gamma-ray burst if it were nearby, within the thickness of the galactic disc, where the required energy might be 10^{37} ergs. It could not explain the superburst in the Large Magellanic Cloud, which emitted 10^{45} ergs, a hundred million times more. It could not explain even the more routine eruptions of the soft gamma repeaters, most of which are several times closer than the Large Magellanic Cloud; these typically involve at least 10^{40} ergs.

For the outburst to be sudden, the infalling object must be dense and solid. A stream of gas would not do. In fact, many neutron stars and black holes are observed that accrete streams of gas, supplied by companion stars in close orbit. These objects are continuous, though flickering, sources of X rays or gamma rays, rather than emitting sudden bursts. The accretion model could only work if large chunks of solid matter were accreted. These would have to be about the size and mass of our Moon to explain superoutbursts, and the size of small asteroids to explain the more frequent repetitive events.

The solar system has nine planets (or eight, for some astronomers consider Pluto too small to merit the dignity of planethood) and thousands of asteroids, rocky bodies ranging from less than a mile to a few hundred miles in diameter, mostly in orbits between those of Mars and Jupiter. However, none of them has fallen into the Sun in recorded history, or will in the future, as far as we understand the mechanics of their orbits.

The reason planets don't fall into the Sun, or the Moon into Earth, is described by a quantity called angular momentum, which measures the extent to which an object attracted by the Sun (or another mass) moves crosswise, rather than directly toward the center of attraction. The planets have a great deal of angular momentum, so they move in nearly circular orbits around the Sun, even though its gravity pulls them directly toward it, nearly perpendicular to their actual motion. Comets have much less angular momentum, so their orbits point almost directly toward the Sun, and a few have actually fallen into it and been swallowed up. Most important, angular momentum is conserved, meaning that it is not changed by the attraction of the Sun (or other central object). A body with high angular momentum keeps it forever, unless a third body intervenes.

In order to explain soft gamma repeaters as resulting from the accretion of solid objects by a neutron star (a much smaller target than the Sun), the rate of accretion must be increased enormously over that observed in the solar system. Somehow, some of the solid objects must lose their angular momentum. We now enter the world of speculative astrophysical model building, in which each step is fraught with uncertainty. The laws of physics must be obeyed, but we are free to guess at the initial conditions—how nature set its machines in motion—and will try to find guesses that make the model explain the observations.

Every soft gamma repeater that has been observed has been found to be in a young supernova remnant, the remains of a star that exploded within the last few thousand or tens of thousands of years. This is license (in fact, makes it almost mandatory) to consider explanations that depend on the fact that a supernova occurred recently, as astronomers measure time. If the pre-supernova star had planets around it, as many stars (not just our Sun) and even one neutron star have been observed to have, then the supernova explosion would have changed their orbits. In general, it would not have destroyed the planets, and if the pre-supernova star lost less than half its mass in the explosion the planets would have remained in orbit.

If the planetary orbits had originally been nearly circular, as in our solar system, then after the supernova they would be strongly elliptical. If there were several planets, some of their new orbits might well intersect. Collisions would occur only when two planets were at the intersection of their orbits *at the same time*, an infrequent event that typically would first happen thousands of years after the supernova. Collisions produce a shower of fragments, and change the angular momentum of the fragments in unpredictable ways. Some of the fragments could be left with so little angular momentum that they would fall onto the neutron star or black hole left by the supernova, making a soft gamma repeater.

There are analogous processes in our solar system, in which asteroids collide with each other and produce streams of solid fragments. In addition, the remains of disintegrating comets produce episodic meteor showers reminiscent of Figure 10-1. The model of a planetary collision followed by accretion has some precedent, and each portion of it appears to make physical sense, even semiquantitatively, but it depends on a chain of hypotheses, each of which is quite uncertain.

The second class of models of soft gamma repeaters also began as an early model for classical gamma-ray bursts. It was the release of magnetic energy of a neutron star, a solar flare writ large. A neutron star was assumed, partly because it may have more magnetic energy to release than any other compact astronomical object, and partly because the 8-second period of the March 5, 1979, superburst fairly cried out (at least to astronomers) "neutron star."

Simply substituting numbers into a well-known formula showed that to power a single superburst would demand a magnetic field of nearly 10^{14} gauss, several times greater than that measured in any other neutron star (or anywhere else in the universe). Only one superburst had been observed, but astronomers had only had orbiting instruments capable of observing them since the Vela satellites were launched in the 1960s. If a soft gamma repeater makes a superburst every few decades, and is active for several thousand years (the age of the supernova remnants in which they are found), then it must make a hundred or more superbursts during its lifetime. The required magnetic field then approaches 10^{15} gauss (magnetic energy is proportional to the square of the magnetic field, so that to increase the energy a hundredfold only requires increasing the field tenfold). There was no direct evidence for this—it is possible that soft gamma repeaters only produce one superburst in their lifetimes and are active only for a few decades—but it was an intriguing possibility.

These dismayingly (or excitingly, depending on one's temperament) large fields demanded experimental verification. In principle, astronomers knew how to do this. The spins of isolated (not accreting gas from a companion star) neutron stars gradually slow. This was discovered in 1968 when the first pulsars were observed, and it proved that pulsars are rotating magnetized neutron stars. Unfortunately, because the slowing is very gradual it was impossible to measure it in the 3 minutes of observation of the superburst of March 5, 1979, and its subsequent outbursts were too brief even to detect its 8-second period.

An indirect argument was possible. An 8-second period was the slowest ever observed for an isolated neutron star. It might have been born with that period, but it might have been born spinning much faster, resembling other isolated pulsars whose spin periods range from 1/30 second to a few seconds. If it was born spinning fast, its magnetic field could be calculated from the rate at which it must have slowed, using a formula first derived for pulsars in 1968. The result was a field of about 6×10^{14} gauss, consistent with the other estimates. This model was later dubbed the "magnetar" (for magnetic star) model, because of the extraordinary magnetic fields that were predicted to be the source of the outbursts.

The difficulty was to test this prediction. This was not done until the late 1990s, when more sensitive X-ray telescopes observed steady radiation from two soft gamma repeaters between outbursts. They were found to have periods of 5.16 and 7.47 seconds, similar to the 8-second period of the repeater in the Large Magellanic Cloud. Sure enough, their spins were slowing at rates that implied magnetic fields in the range 10^{14}–10^{15} gauss, just as predicted.

This apparent triumph of the magnetar model was soon confused by an unexpected discovery. Although the spins of soft gamma repeaters were slowing at rates that implied large magnetic fields, they were slowing unevenly. The theory predicted nearly uniform slowing, as is observed to high accuracy for radio pulsars. This appeared to vitiate the success of magnetar models and point again to accretion models, in which a slow but steady accretion of gas (in addition to the solid chunks required to explain outbursts) would produce irregular variations in the neutron star's spin period, as observed for many neutron stars known to be accreting gas. Neither model was quite satisfactory, but neither was conclusively disproven. The true nature of soft gamma repeaters remains a matter of controversy.

Whatever soft gamma repeaters turned out to be, the path to considering cosmological distances for the classical bursts was now open. No longer would that idea be met with the damning reply that at least one burst was known to come from the Large Magellanic Cloud, almost next door.

Cosmological distances were still not proved correct, and in fact the majority of gamma-ray-burst astronomers still rejected them. There were several reasons for this. The reports of archival visible counterparts and of gamma-ray spectral lines, although never quite

conclusive, were widely accepted; there was no evidence directly con-
tradicting them, and it was natural to seize upon any clues that appeared. The compactness arguments (Chapter 5), together with a general reluctance to claim that gamma-ray bursts were millions of times
more luminous than anything else in the universe (including entire
galaxies and quasars), led to a strong prejudice against cosmological
distances, which seemed to be the most radical hypothesis. Scientists,
even theoretical astrophysicists, learn to temper their speculations
with a dose of conservative skepticism. This is essential to avoid turning the formulation of hypotheses into unbridled and unsubstantiated guesswork. Usually skepticism is justified, although the universe has proved remarkably clever at finding unexpected ways to
release enormous amounts of energy.

In 1986 two Princeton University astrophysicists, Jeremy Goodman and Bohdan Paczyński, who had retained faith (or, at least, interest) in the cosmological distance scale when nearly everyone else
had lost it, reexamined the compactness argument. In independent
but closely related (and contemporaneous) calculations they asked
what would happen if an enormous amount of energy or power, far
exceeding the compactness limit, were released in a small, neutron
star-sized, region of space. The answer was that an expanding gas of
electron–positron pairs and gamma rays would form, and that this
fluid would expand outward at nearly the speed of light. As it expanded it would cool, converting its heat to the energy of outward
motion, just as rising and expanding air in a thunderstorm cools until
rain condenses, and that rain freezes into hail. As it cooled, the energy
of the average particle and the average gamma ray would decrease,
until the gamma rays were no longer energetic enough to make new
electron–positron pairs to replace those lost by annihilating into
gamma rays. Eventually, far from the source, there would only be
gamma rays streaming nearly radially outward.

It was possible to calculate the appearance of such an event. It
would have a Planck (black-body) spectrum with a comparatively narrow spectral peak at a gamma-ray energy of roughly a few MeV. Its
duration and time dependence would be the duration and time dependence of the energy release that powered it. The most obvious way
of making such an incredibly energetic event involves the collision
of two neutron stars, or of a neutron star and a black hole of similar
mass. Such a collision produces a single intense release of energy lasting a fraction of a millisecond. The predicted gamma-ray burst would
consist of a single pulse, just this short. In each respect—spectrum,
duration, and time dependence—these theoretical predictions disa-

greed with the observed properties of classical gamma-ray bursts. The compactness argument, examined closely, did not kill the cosmological hypothesis, but showed that, if it were true, astrophysicists had no inkling of how bursts with the properties we observe are actually made.

11

BATSE

In the 1970s NASA began planning a series of four "Great Observatories." These large orbiting telescopes were meant to move astronomy from the era of observing on the ground, hampered by Earth's atmosphere, to a new age in space. The glories of the universe revealed by space astronomy would keep NASA in the public eye, and its budget in the good graces of Congress.

It came half true. The most famous Great Observatory, the Hubble Space Telescope, was finally launched in 1990, several years late. An error in its optics initially made it nearly useless. However, this error was fixed by the insertion of a correcting mirror, and the Hubble Space Telescope has fulfilled most of the dreams of NASA's publicists, and even of astronomers. Although it made only a small contribution to the most fundamental astronomical advance of that period, the discovery of the acceleration of the cosmic expansion (Einstein's cosmological constant), the Hubble Space Telescope was used to make many other remarkable discoveries that could not have been made from the ground because of Earth's blurring blanket of air. Its spectacular photographs are the pride of every glossy popular astronomy book.

The other three Great Observatories are better known to professionals than to the general public. The Gamma-Ray Observatory (GRO) was launched April 5, 1991. It required an emergency space walk from the shuttle that launched it to deploy a stuck antenna (necessary for sending data back to Earth), but after that it supplied data until 2000. BATSE (the Burst and Transient Source Experiment) was part of GRO. A large X-ray telescope (Chandra) was successfully launched in 1999. An infrared telescope, far behind schedule, is scheduled for launch in 2002. The delay of these last two telescopes meant that the hope that the four Great Observatories would complement each other with simultaneous observations of the same objects would not be fulfilled.

Each of these Great Observatories was more than a decade in the making, was launched far behind schedule, and was substantially smaller and less capable than originally planned. Several factors contributed. NASA's imprudent dependence on the space shuttle meant that the *Challenger* explosion in 1986 delayed all launches. But the problems were deeper. NASA is unique among U.S. government agencies in not having a well-defined and generally accepted mission, and has not had one since the first men were landed on the Moon in 1969. A mission, such as defending the country or providing medical care to old people, even if inefficiently pursued, provides a lodestar to keep an organization on course. In contrast, NASA never seems to know what it should be doing, or why. It has invented for itself a series of large but pointless projects (space shuttle, space station) whose chief function appears to be to keep the budget up for as long as possible. Science is only decorative, a means to raise itself in the public's (and Congress's) estimation.

The practical effect is a perpetual Ponzi scheme of promises. Great things are advertised, for fear that a more modest and realistic proposal would not be funded at all, just as a Ponzi scheme promises investors returns too good to be true. Without a clear mission there is no definition of what is worth doing, how much of it needs be done, or even of success.

Promises are cheap, but realizing them is expensive, and the budget is never sufficient because the promises are inflated. They must be shrunk, and completion is delayed many years, sometimes until their rationale is gone. Delay and redesign are themselves expensive. The final cost is much greater and the final capability much less than if the original promises and budget had been realistic.

Many NASA projects have suffered. The worst offenders are the shuttle and space station, whose enormous appetite for funds and top priority have squeezed everything else NASA does. It is in NASA's interest to plan projects that drag on as long as possible, for that assures stable funding and defers the question of why the organization exists at all. Delay is not a bad thing for bureaucrats. Once a great deal of money has been spent, outright cancellation is unlikely.

Despite this, scientists have done excellent work with what NASA has given them. BATSE was developed under the leadership of Gerald Fishman. A member of an old St. Louis family, his entire career has been spent in Huntsville, Alabama, working in X-ray and gamma-ray astronomy, first for a NASA contractor and then directly for NASA. Driving a pickup truck with a gun rack, he passes for a southern "good old boy" until he starts talking about gamma-ray bursts in standard mid-American English. The empty gun rack (when I saw it) may also betray him.

BATSE, planned and eagerly awaited by gamma-ray-burst astronomers since the 1970s, finally was launched as part of the Gamma-Ray Observatory more than a decade later. BATSE had eight flat laundry-basket-sized (20 inches in diameter and half an inch thick) NaI(Tl) scintillator detectors mounted as if on the faces of a regular octahedron. This design had been carefully developed to resolve the Copernican dilemma and answer the question of the distances of the gamma-ray bursts. GRO carried three additional instruments to observe higher-energy gamma rays (scintillators are ineffective at energies of more than about 2 MeV). These instruments would collect data from gamma-ray bursts, as well as observe other astronomical objects. In fact, BATSE was a comparatively small and cheap part of GRO.

Fishman's strategy consisted of two parts. The first was to collect data on a large number of bursts, both strong and weak, to determine whether the deviation of the $-\frac{3}{2}$ power law relation between the number of counts N and their brightness S was real. Most previous experiments had used small orbiting detectors that observed some of the rare strong bursts because they collected data for a long time, but were too small to detect weak bursts. The remaining experiments, BATSE prototype detectors flown on balloons by Fishman's group, had collected too few data (because balloon flights were limited to about a day's duration) to see any of the rare strong bursts at all. As a result, the apparent deficiency of weak bursts detected by the balloon flights, compared to those expected from the insensitive orbiting detectors, might be attributed to errors in relative calibration of the instruments. BATSE, using large sensitive detectors to be exposed to the sky for years, promised to obtain a homogeneous set of data including both strong and weak bursts that would settle this question definitively.

The second part of the BATSE strategy was an attack on the problem of the angular distribution of the bursts on the sky. Most astronomers had thought that the way to solve the gamma-ray-burst problem was to use an interplanetary network of detectors to obtain very accurate positions of a few bursts, positions that could then be searched for visible counterparts. This was the same method used to analyze the Vela data to show that the bursts did not come from Earth or the Sun, but improved because the satellites in the interplanetary network were much farther apart than the Vela satellites had been. Unfortunately, data analysis was still slow, so the positions only became available long after the bursts. This approach, tried since the late 1970s, had invariably failed. Any counterparts must either be very faint or fade quickly.

Fishman proposed to obtain a large collection of very approximate positions using a method previously employed by Mazets. Any gamma-ray burst would be observed by several of BATSE's eight detectors. Because the detectors were all on the same spacecraft, they would detect the burst at the same time, but would measure different intensities. The signal would be strongest in a detector oriented perpendicular to the direction to the burst, and weakest in a detector edge-on to the burst. This is analogous to the fact that the Sun casts its brightest light and most intense warmth when it is nearly overhead in the sky, at midday in summer, and very little when it is near the horizon. From the ratios of the strengths of the signals in the detectors the direction to the burst could be derived.

The resulting angular positions of the bursts would be rather approximate, partly because of statistical fluctuations in the counts of gamma rays, and partly because of uncertainties in the detector calibrations. The expected error would be about 4 degrees of arc (eight times the apparent diameter of the Sun and Moon) for the strongest bursts, and about 10 degrees (about the apparent size of your fist with your arm extended) for the weakest bursts, for which the statistical errors are disproportionately large. There would be no hope of finding visible counterparts, but there would be a definitive answer to the question of the broad distribution of bursts on the sky: Are they concentrated toward the galactic plane, toward the center of our galaxy, or are they distributed uniformly on the sky? Obtaining a large homogeneous body of data was essential, but precision in the individual positions was not.

BATSE succeeded in each of its objectives. Eight months after launch the Huntsville group published its first results, based on the initial 6 months of data. Figure 11-1 shows the number of bursts whose peak count rate exceeds a value C_{max} (equivalent to the brightness S discussed in Chapter 9) for a range of values of C_{max}. More bursts exceed small C_{max} than exceed large C_{max}, but the deficiency of faint bursts, compared to the expected $-\frac{3}{2}$ power law, is unquestionable.

Figure 11-2 shows the distribution of the 153 bursts on the sky, plotted in galactic coordinates. It is evident that there is no concentration toward, or away from the galactic plane or the galactic center,

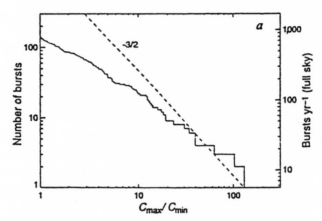

Figure 11-1. The number of bursts N with a peak gamma-ray-count rate (equivalent to the brightness S) exceeding C_{max} vs. C_{max} out of 140 gamma-ray bursts observed by BATSE in its first 6 months. The brightnesses are measured in units of the minimum count rate C_{min} below which the signal is not recognized as a burst at all. The dashed line gives the relation N proportional to C_{max} (or S) predicted if space is uniformly filled with burst sources. The deficiency of faint bursts is evident. (*Nature* V. 355, p. 144 [1992].)

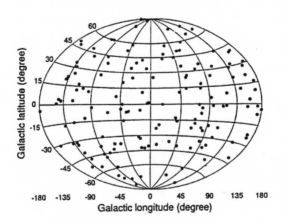

Figure 11-2. The distribution of gamma-ray bursts on the sky, in galactic coordinates, as observed by BATSE in its first 6 months. The distribution is isotropic, with no evidence for a concentration in the galactic equator (galactic plane or disc), toward the galactic center, or in any other direction. All apparent clustering is satisfactorily explained as a result of statistical fluctuations (counting statistics). Compare to Figure 11-3. (*Nature* V. 355, p. 144 [1992].)

conclusions confirmed by quantitative statistical analysis. The apparent nonuniformities, such as the absence of bursts between $-45°$ and $-60°$ galactic latitude and $-90°$ and $+90$ galactic longitude, were all satisfactorily explained as the result of statistical fluctuations, and disappeared with additional observations. Through its 9-year life BATSE detected nearly 3000 bursts, and only reconfirmed these conclusions with ever-increasing accuracy.

No longer could astronomers hope that the Copernican dilemma would disappear with improved data. The data were in hand, and their implication inescapable: we are at the center of a spherically symmetric distribution of gamma-ray-burst sources, and this distribution has an outer edge. Beyond this edge the density of burst sources decreases to insignificance.

Only two plausible distributions of sources meet this requirement. One is a spherical cloud, centered on the center of our galaxy. We would not be exactly at the center of such a cloud because we are not at the center of the galaxy, so we should see a greater depth of cloud, and more gamma-ray bursts, in the direction toward the galactic center than in the opposite direction. If the cloud is very large, this difference would be small, perhaps small enough to be consistent with the data in which no such difference is apparent.

An extended cloud or halo around the galaxy has been the subject of speculation for many years because of the "missing mass" problem. The universe contains at least ten times as much mass as is observed in stars, a fact discovered by the Swiss astronomer Fritz Zwicky in the 1930s. The nature of this matter is still completely unknown. It might be black holes, neutron stars, very dim stars, or dark planet-like objects between the stars or, conceivably, exotic elementary particles of varieties as yet undiscovered. Galaxies themselves contain "missing mass," for their outer parts revolve around their centers too fast for the gravity of their known stars to hold them together. Additional mass must be present. The spatial distribution of this mass is unknown in detail, but it must extend to enormous distances (many times the 30,000 light-years between Earth and the galactic center). These facts, although scant, entitle the astronomer to speculate about a very large spherical halo to our galaxy, with us comparatively close to its center.

The distribution of gamma-ray bursts should be compared to that of pulsars, which are known to be magnetic neutron stars in our galaxy. This is shown in Figure 11-3. It is evident that the pulsars are strongly concentrated in the galactic plane. Within that plane, they are much more likely to be found in the hemisphere toward the ga-

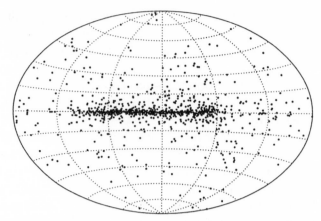

Figure 11-3. The distribution of radio pulsars (which are magnetic neutron stars) on the sky, in galactic coordinates. A concentration in the galactic equator and generally toward the galactic center is obvious. This is a complete catalogue, collected from many sources, so the sensitivity of the surveys varies somewhat with direction, but most of the concentration is real rather than an artifact. (D. R. Lorimer.)

lactic center than in the opposite hemisphere. This is entirely unlike the isotropic distribution of gamma-ray bursts. If bursts are produced by magnetic neutron stars, their origin must be different from that of the other magnetic neutron stars we know, the pulsars. If born in the same places as pulsars, burst sources must escape at high speed, equally in all directions. Even this is not sufficient, because bursts produced too soon after their sources begin moving will be concentrated toward the galactic plane and center. These assumptions are not demonstrably impossible, but they do strain credulity.

The alternative distribution of burst sources is that originally suggested by Usov—uniformly in the universe. Their isotropy then follows from the universally accepted fact that on large scales the matter in the universe is isotropically distributed around us. The deficiency of faint bursts could readily be explained either by the cosmological redshift of distant objects (which also makes them appear fainter and harder to detect) or by an intrinsic difference between the younger and more distant regions of the universe and those near us; in the remote past the universe might have produced fewer bursts per galaxy. The redshift effect was certainly true, and the latter was plausible, though unproven. Either was sufficient.

Astrophysicists would now have to face the task of explaining gamma-ray bursts so distant that they radiate roughly 10^{52} ergs of energy in gamma rays, and do so in a minute or less. This is roughly ten times the energy and ten million times the power radiated by a bright supernova, and requires the annihilation of about ½₀₀ of a solar mass of matter into gamma rays. The total energy release was not the difficulty—the formation of a neutron star in a supernova was calculated to emit about ten times this much energy, in a similar time, into neutrinos.* This prediction was verified directly when neutrinos from Supernova 1987A in the Large Magellanic Cloud were observed by particle-physics experiments deep underground. The difficulty was explaining how this energy appeared as the gamma-ray bursts we observe. Many astrophysicists shrank from the challenge.

BATSE continued to collect data for 9 years. No satellite lives forever. Something breaks, or something else is used up. If the satellite is in a low orbit, the friction of the uppermost fringes of Earth's atmosphere slowly bleeds away its angular momentum and its orbit shrinks, until it finally burns up in the atmosphere.

The Gamma-Ray Observatory was stabilized by gyroscopes, essential components of most spacecraft guidance systems. It had three. At the end of 1999 one failed. GRO could be stabilized with two, but if one or both of the remaining two were to fail suddenly and unexpectedly, there was concern it might tumble uncontrollably.

This would certainly have meant the end of science from GRO, but it was an unknown time, perhaps many years, away. Yet NASA decided deliberately to bring down the observatory to burn up over the Pacific Ocean. On June 4, 2000, NASA reported the end of GRO under the euphemistic headline "Compton Gamma-Ray Observatory Safely Returns to Earth."

Why? NASA's explanation was that if control were lost and the satellite were to tumble, its atmospheric reentry (which would surely happen in the next decade as a result of atmospheric friction) would be unpredictable and uncontrollable. Some of its heavier pieces, surviving reentry, might be hazardous. The uncontrolled reentry of the first American space station, Skylab, in 1975 caused a great deal of excitement and scattered debris over Australia, but damaged nothing and harmed no one. Estimates of the chances that someone, somewhere would be killed by an uncontrolled reentry of GRO, in which pieces could land almost anywhere on Earth (except at high latitudes), ranged from 1 in 1000 to 1 in 10,000.

This risk is small, but not zero. It may be too much. In fact, the

---

*Neutrinos are particles, produced in certain nuclear reactions, that interact so weakly with matter they pass freely through Earth and can carry energy out from the dense interior of neutron stars

approximately one hundred thousand employees of NASA and its contractors subject the rest of us to greater risk each day they drive to work. Perhaps, if there were an uncontrolled reentry, NASA should tell its people to stay home for a day to make up for it. However, NASA engineers had devised a procedure to control reentry, so it would occur safely over empty ocean, even with no functioning gyroscopes. There really was no estimatable risk at all. Despite this, NASA management ignored both the advice of its engineers and the pleas of the scientists and ordered the destruction of GRO. Why?

NASA has a long history of launching scientific satellites and then losing interest in them. The glamour is in the launch, not in years of painstaking research, especially when it does not produce color pictures. Many scientists believed NASA seized on the gyroscope failure as an excuse to stop paying for the operation of GRO and the analysis of its data.

The truth was more devious. NASA, and especially its administrator, Dan Goldin, were fixated on the space station. This was an international enterprise, in which Russia played a significant part. Yet the Russians were still committed to *Mir*, their own troubled and deteriorating space station, which had been in orbit since 1986, and this commitment interfered with their contribution to the joint enterprise. NASA was determined that the Russians de-orbit *Mir*. The failure of a gyroscope on GRO offered an opportunity: NASA would make this the excuse for sacrificing GRO, otherwise healthy, to set an example for the Russians. It was cynical, but it worked; early in 2001 the Russians de-orbited *Mir*.

For the scientists working on BATSE, and the other GRO instruments, it was the end of a dream. For many, it was also the end of their careers. BATSE had occupied a quarter of a century, from the first balloon flights of its prototypes to its "safe return to Earth." Jerry Fishman, its prime mover, was 57 when it burned up. He had devoted nearly his entire scientific career to this one experiment. Who would appoint a man of 57 to lead a new project that also might last decades?

Large space experiments cannot be launched often. They are expensive (GRO cost about a billion dollars, and the Space Telescope several billion), and only a fantasist would suggest that the United States, or the world, would build several every year. Scientists who work on them accept that they will be involved with very few experiments in their careers. Sometimes, because of cancellations or failures, there never are any data. These people must think of themselves more as the builders of permanent facilities, such as giant ground-based telescopes or particle accelerators, which operate for decades, than as experimental scientists in a laboratory, who may prepare and conduct an experiment in days, weeks, months, or, at most, a few years.

Space astronomy is a unique kind of science. It is a blend of traditional astronomy (evoked by NASA's term "Great Observatories"), in which an observatory and its instruments accommodate many users who observe many different objects, and laboratory science, in which a single experiment is planned and built to answer a single question. Often, it seems to have all the disadvantages of each—only a single question is answered, but no new instrument will be available for a very long time. Still, BATSE was a grand success.

12

The Great Debate

Science is full of controversies. One may even say it is about controversy, because experiments are usually performed to decide which of two (or more) proposed views of the world is correct.

The BATSE results tilted the scales toward cosmological distances for gamma-ray bursts, but not all astronomers were convinced. Disagreements about astronomical distance scales were not new. The most famous such controversy had occurred early in the twentieth century, and had two parts. One concerned the size of the Milky Way, and how far we are from its center. The second, and more fundamental, concerned the spiral nebulas, which we now know as galaxies. It was then uncertain whether these nebulas (so called because they appeared fuzzy, or nebulous, through a telescope) were galaxies resembling our Milky Way (then termed "island universes," although no one maintained that they were actual universes disconnected from our own), or whether they were simply clouds of gas within our own galaxy, illuminated by stars within them. Many other nebulas were known to be such clouds.

In 1920 the National Academy of Sciences, the chief scientific honorary society in the United States, invited two eminent astronomers, Heber Curtis and Harlow Shapley, to discuss and debate "the distance scale of the Universe." Shapley is better remembered today, partly because he outlived Curtis by 30 years, and partly because, as the author of a number of successful popularizations and director of the Harvard College Observatory, he later became a prominent spokesman for astronomy as a whole. Curtis made a comparatively small estimate for the size of the galaxy, and Shapley a comparatively large one, at least sixfold greater than Curtis's. Neither was close to the best modern values, which lie in between.

Curtis argued that the spiral nebulas are galaxies like our own, while Shapley maintained the contrary. Shapley's strongest evidence lay in measurements made by a Dutch astronomer, Adriaan van Maanen. Comparing photographs taken decades apart, van Maanen had found that the arms of some spiral nebulas appeared to be rotating. If true, this would have required that they be comparatively close to us, within our galaxy, for at truly extragalactic distances any apparent rotation would have been too slow to observe.

Like all scientific disagreements, the debate was settled not by persuasion in the lecture hall but by new data. In 1924 Edwin Hubble discovered Cepheid variable stars, astronomers' favorite standard candles, in the Andromeda nebula. These showed that this nebula was really an "island universe," a galaxy like our own but far outside it. Curtis was proved right by Hubble on this most important issue, ironically by a method that Shapley had pioneered. Hubble went on to discover the expansion of the universe, today known as the Hubble law, the most fundamental fact in cosmology.

Remeasurement of van Maanen's photographs showed that he had erred; no rotation of the arms of the nebulas was measurable. Why he erred was never established. He was Shapley's friend; unintended bias is only human and nearly unavoidable. Shapley himself said, "I believed in van Maanen's results . . . after all, he was my friend."

The debate itself was little noted at the time. It was remembered in legend as the study of galaxies and cosmology became larger and more important parts of astronomy, especially in the mind of the lay public. It has been recounted in several popular and historical books, and is now generally referred to as the "Great Debate."

In April 1995, 75 years (to the week) after the "Great Debate," another debate entitled "The Distance Scale to Gamma-Ray Bursts" was held in the same room. The issue was whether gamma-ray bursts originate in an extended halo of our galaxy or at cosmological distances, the two possibilities permitted by the BATSE data.

The organizers' most important task was to choose protagonists. This was not as simple as might be thought. In 1920 theoretical astrophysics hardly existed, and both Curtis and Shapley were observational astronomers who themselves had obtained most of the data that they used in their arguments. Seventy-five years later theoretical astrophysics was a busy specialty, and theorists had taken the initiative in interpreting many astronomical discoveries. The observers were often in the background. This was particularly true in gamma-ray astronomy, most of whose observers came from experimental physics rather than classical astronomy. Jerry Fishman presented an overview of gamma-ray-burst astronomy at the 1995 debate, chiefly based on data from BATSE, but all the other speakers were theorists.

The pool of possible protagonists was surprisingly small. Difficulties in obtaining research grants had driven many American theorists out of the field. Although theoretical work is cheap, requiring no special equipment or laboratory space, it was unpopular at the two U.S. government agencies that support astronomy. The National Science Foundation,* beginning in the 1980s, withdrew most of its support of theoretical astrophysics. NASA has a small program in astrophysical theory, but it was badly administered. Its managers took pride in keeping the fraction of proposals approved very low, around one in six. This enabled them to boast to their superiors that they supported "only the best."

It is impossible to predict how good a scientific idea will be before it has been tried. With no certain basis for judgment, grant selection, based on the ratings of a panel of reviewers, turns into a popularity contest. Most proposal reviewers are themselves recipients of grants, usually in more established fields, and it is easy for them to turn into a closed club, admitting no new people or ideas. There is a built-in bias against innovation and originality, because new ideas are unproven and vulnerable. In a controversial and unsettled field like gamma-ray bursts, in which even basic assumptions were open to dispute, no proposal will satisfy all reviewers, and all are likely to be rated poorly. Responsible leadership would have managed the selection process so that work on new ideas and in controversial fields could be supported, but this was not done.

The level of acrimony was shown by the XYZ Affair. The initial BATSE results led to a competitive frenzy of interpretation by theorists. X submitted a paper to the *Astrophysical Journal,* the principal technical astronomical publication in America. After several months' delay he received an anonymous referee's report† describing his work as "a modest rehash of the work of Y and Z" (Y and Z were frequent

---

*One very prominent observational astronomer, discussing the difficulty of finding funding for new ideas, referred to "NSF's failure to perform a useful social function."

†Journal editors send submitted papers to reviewers whose identity is known only to the editor. The reviewers (also called referees) advise the editor whether to accept or reject the paper or require the author to change it. This process is called "peer review" and is meant to assure the reader that published results are valid.

collaborators), along with other remarks in a similar tone, all unsubstantiated. This came with a form letter of rejection from the editor.*

So far, this was just ordinary warfare among the theorists, perhaps a bit cruder than usual, designed to delay publication in an attempt to reduce the credit and priority that would be given the author if his ideas turned out to be successful. All parties knew that almost any theoretical paper, good, bad, or indifferent, that was not actually crackpot would be published eventually, somewhere. In the meantime, however, someone else (conceivably the reviewer) might publish the idea first, or the idea might diffuse into the general consciousness, reducing the paper's impact.

The really devious trick was that while the referee's report had been produced on a computer printer, the name of Y had been added later, apparently with a typewriter, on top of an erased space just wide enough to accommodate the word "myself" in the original. It looked as if the report had been written by Y who had thoughtlessly written "myself and Z," and who at the last minute had sought to maintain anonymity by erasing the word "myself" and replacing it with his actual name, as someone else would have written.

The effect was to infuriate X, and to persuade him that his competitor Y had written a grossly unfair evaluation of his work and had tried to conceal his identity in a stupidly incompetent way. X wrote to Y (and to Z) accusing him of this, and thereby destroyed what had been a cordial, though competitive, relationship. Y denied being the referee, and the editor (in a rare relaxation of secrecy) confirmed this. Apologies from X to Y and Z did not heal the breach. The guilty party (X has suspicions based on other internal evidence, but learned not to jump to conclusions) thus succeeded not only in delaying a timely paper by several months, but in sowing widespread dissension and distrust. In this environment, which extended far beyond the participants in this affair, it was hardly surprising that, year after year, not a single gamma-ray-burst theory proposal received the consensus support needed for approval.

Without a research grant a scientist at an American university cannot travel to meetings or publish in mainstream scientific journals

---

*Editors of astronomical journals have generally followed referees' advice blindly, so that even a plainly baseless negative review triggers an automatic letter of rejection. Experienced authors know they can request a new referee and start over. Of course, once this has happened to someone, he is eager to exact revenge the next time he is a referee himself. He does not know who the guilty party was, though he has probably spent some time guessing, so anyone may be the next victim.

(whose costs are paid by a fee levied on contributors). Universities expect these expenses, as well as the actual costs of research, to be paid by outside grants. Funding is the tail that wags the dog of research; no grants mean no research. Many scientists complain that they spend as much effort writing grant proposals as actually doing the work. The result of the NSF's and NASA's policies was an exodus from theoretical astrophysics, and especially from work on the more controversial topics like gamma-ray bursts. Astrophysicists moved to Los Alamos to work on nuclear weapons, changed their research to less controversial (and less interesting) parts of astrophysics, or switched to other fields. Leadership fell into the hands of scientists abroad, in Britain, continental Europe, and Israel.

The organizers of a meeting at the U.S. National Academy of Sciences wanted American speakers. By 1995 cosmological distances were nearly a consensus position, so they had little trouble finding a prominent scientist to argue for them. They chose Bohdan Paczyński of Princeton. Born in Poland, he came to America early in his career. He is best known for his work on gravitational microlensing, a phenomenon in which the gravity of one mass (a faint star, or a black hole, for example) focuses the light of another object directly behind it onto an observer, making the more distant object appear much brighter than it would otherwise. Microlensing was first suggested in an attempt to explain the extraordinary brightness of quasars. Investigated quantitatively, this hypothesis failed because, to produce a significant brightening, the light source, the lensing mass, and the observer must be lined up extremely accurately, which is rare.

The theory of gravitational microlensing was worked out in the mid-1960s by the Norwegian astrophysicist Sjur Refsdahl. Twenty years later Paczyński pointed out that with modern electronic detectors that could measure the brightness of many thousands of stars, night after night, even rare microlensing events might be detected. Several experiments now observe dim masses in the halo of our galaxy, otherwise too faint to detect at all, when they pass in front of and focus the light from more distant but brighter stars. Paczyński was also one of the very few astronomers who consistently backed the extragalactic distance scale for gamma-ray bursts when it was unpopular, and therefore was a natural choice to speak on its behalf after it became the consensus view.

For a speaker against cosmological distances the organizers chose Donald Lamb of the University of Chicago. They had few options, for by 1995 he was the only prominent scientist who still held this position. In 1993 Lamb and a student, Jean Quashnock, had published two controversial papers in its favor. In the first they had examined the BATSE data and reported finding evidence of a small, but statistically significant, anisotropy in the distribution of bursts on the sky.

Why did their result differ from everyone else's? There is a key sentence in their abstract (a short summary presented before the main text of almost every scientific paper): "We show that the 54 Type I . . . bursts lying in the middle brightness range. . . ." Instead of considering all the bursts in the available BATSE data, they picked and chose. They defined a measure of variability, divided the bursts into two classes (Types I and II) on the basis of this measure, and selected only one class. They defined a measure of brightness, divided the bursts into three classes on the basis of this measure, and considered only one class. Combining these, there were six subclasses of bursts they could consider. They only reported on one subclass, the one for which they found evidence of anisotropy.

It was even worse. Their choices of variability and brightness classes were themselves arbitrary. They might have chosen different boundaries between the classes, or even used other variables (such as the spectrum) to define classes. We do not know how many different possible selections and subdivisions they tried, searching for anisotropy, either explicitly (by actually performing the calculation) or implicitly (by looking at the data, in effect, making a rough mental calculation), but surely many more than six.

How significant was their result? They quoted a deviation from isotropy of 2.6 standard deviations (usually referred to by the Greek letter "sigma"). The standard deviation is a measure of the purely statistical scatter of a measurement from its expected value, given the assumptions made. If the measurement is simply a count of the number of times something happens, the standard deviation is the square root of the number expected. For example, for a fair coin tossed 100 times the expected number of heads is 50, and the standard deviation of the number of heads is nearly 7, 14% of the number expected. The standard deviation is larger if the count is higher, but it is also a decreasing fraction of the count for larger counts. With 10,000 tosses 5000 heads are expected, with a standard deviation of 71, only 1.4% of the expected number.

If you toss a fair coin 10,000 times, there is a 68% probability that the actual number of heads will lie in the range 4929 to 5071, differing from the expected number of 5000 by no more than one sigma, or 71. The deviation is less than two sigma (the range 4859 to 5141) 95% of the time, and it is less than three sigma (the range 4788 to 5212) 99.7% of the time.

A "three sigma" or greater result sounds very significant; it has only about a 0.3% chance of occurring as a result of statistical fluctuations alone. In an ideal experiment it would be convincing evidence that you had made a valid observation. In a series of coin tosses it would probably persuade you that the coin was not fair.

Unfortunately, real experiments are far from ideal. There are often systematic errors, not described by the laws of statistics. For example, a person tossing a coin may start with it heads-up more often than with tails-up. This hardly matters if it tumbles randomly many times before coming to rest. However, it is possible, deliberately or inadvertently, to toss a coin so that it spins stably without tumbling and lands with the same side up as that with which it began. In that case the final count of heads would not tell you whether the coin was fair because the tossing was not fair. If an unfair coin were used as a source of random numbers in some other experiment, that itself would be a source of systematic error.

Real experiments are much more complicated than coin tosses, and systematic errors often creep in, despite the experimenter's best efforts and without anyone's realizing it. Such prosaic problems as variations in temperature, humidity, and electrical supply voltage all may affect the properties of apparatus and be sources of systematic error. Physicists have a proverb, learned by hard experience: Half of all three sigma results are wrong. To be convincing a result must stand out of the data "like a sore thumb," like the anisotropy of the pulsar distribution shown in Figure 11–3. That is surely no statistical fluke, and it is not a figment of systematic error. More cynical physicists have another proverb, attributed to the statistical mechanician Paul Ehrenfest: if it is essential to use probability to prove that you are right, you are usually wrong.

Lamb and Quashnock's result was only 2.6 sigma. If there were no real anisotropy in the distribution of burst sources, there would be about 1 chance in 100 of finding this result as a result of statistical fluctuations alone. However, they found anisotropy in only one of six subclasses of the data. That is like trying the experiment six times and succeeding once; the possibility of it happening by statistical fluctuation is about 6 chances in 100. The vast universe of other possible criteria to select data that they might have used, but that did not show any anisotropy, eroded the significance of their result even further. And, of course, there was the possibility of systematic error. Few were convinced.

Even "sore thumb" results may be the result of systematic error. Systematic error is a difficult problem because (by definition) it is not understood, and hence its size and properties are unknown. If it were understood, it would be eliminated (e.g., by ensuring that a tossed coin tumbles, or by beginning half the tosses with it heads-up, and half tails-up). The presence of systematic error is not necessarily the fault of the experimenter. It is unavoidable in the messy world of real instruments, whose properties and foibles are never known perfectly.

The best way to deal with systematic error is to repeat the experiment independently, using methods as different as possible from those of the original experiment. A new experiment should have smaller, or at least different, systematic errors. This could not be done with the gamma-ray-burst data, for no one was going to build a duplicate BATSE. However, the data could be reanalyzed, which may be almost as effective.

Immediately following Quashnock and Lamb's paper was a paper by Robert Rutledge and Walter Lewin of MIT that did exactly that. Quashnock and Lamb had selected bursts on the basis of their brightness as measured by count rates in the BATSE detectors. Rutledge and Lewin noted that, depending on its direction and the pointing of the spacecraft, a given burst might produce a greater or lesser signal in the detectors. It is straightforward to correct for this to find the count rate that each burst would have produced if one of the eight BATSE detectors had been pointing directly toward it. If bursts of a certain range of brightness are concentrated on the sky, surely it must be the actual brightness of the bursts that matters, not the apparent brightness measured by a detector whose sensitivity depended on where it happened to be pointing during the burst. Making this correction, Rutledge and Lewin repeated Quashnock and Lamb's analysis and found no significant anisotropy.

Next in the journal was a second paper by Quashnock and Lamb, in which they claimed to find evidence in the BATSE data that gamma-ray bursts repeat. If true, this would be a strong objection to cosmological distances because at such great distances bursts must be so energetic they can only be catastrophic events, destroying the stars that make them.

If the position of a gamma-ray burst is accurately measured, then it is simple to decide if it has erupted again. However, BATSE only gave approximate positions, with errors of several degrees of arc. As a result, it is easy for the measured positions of bursts from two different places on the sky to agree to within the crude accuracy of measurement, while a genuine repeater will generally be measured to have different positions at each eruption. The only way to solve this problem is to look for a statistical tendency of measured burst positions to bunch together in clusters, rather than be scattered uniformly on the sky. If a single burst source erupted a hundred times (like one of the soft gamma repeaters), this clustering would be obvious, but if a fraction of the bursts repeated only once or twice, the clustering would be subtle, and could only be found by statistical analysis. Quashnock and Lamb claimed to find clustering, and concluded "that 'classical' gamma-ray burst sources repeat on a time-scale of months." This conclusion was similar to that implied by the reported archival visible counterparts to bursts (Chapter 8).

This paper was immediately followed by another, by Ramesh Narayan of Harvard and Tsvi Piran of the Hebrew University, disputing its conclusion. Recognizing the possibility of systematic errors in the BATSE positions of bursts that might produce spurious clustering, they performed a simple "numerical experiment." They looked for a correlation between bursts at antipodal (exactly opposite) positions on the sky, such as between the North and South Poles. Sure enough, they found it. Clearly, it could not be real; there is no conceivable way a burst in one direction could be connected with one in the opposite direction, however distant (or close) they are. The inescapable conclusion was that a subtle systematic error in the BATSE burst positions was responsible. This discredited all claims based on such correlations, including those of Quashnock and Lamb.

After this debacle few took seriously the possibility that gamma-ray bursts originate in a galactic halo. Further BATSE data continued to arrive, and the evidence for cosmological distances only grew stronger. Even the Chicago group eventually retreated to the position that only some fraction of the burst sources were found in a galactic halo. This could never be disproved, for, if the fraction were small enough, its contribution to the distribution of bursts would always be less than the uncertainty in the data. Such an untestable (philosophers would call it unfalsifiable) hypothesis is hardly a contribution to science. Like saying there is an alligator under your bed that hides whenever you look, it belongs to another class of thought entirely.

Debates do not settle scientific disagreements, but they air them and focus attention on the issues. The legacy of the 1920 Great Debate was a boom in professional and public interest in its subject, and the organizers of its sequel probably hoped for a similar boost. The purpose of a public debate, like that of awarding a prize, is publicity. Each may be a newsworthy event, while the day-to-day progress of science, immersed in doubt and uncertainty, is usually appreciated only by the participants.

The debate of 1995 was an anticlimax. The protagonists restated positions in a controversy that most astronomers had considered settled in 1993. In the two intervening years, and subsequently, the case for a galactic halo of burst sources had only weakened further. The Great Debate of 1920 concerned active areas of scientific controversy, with strong evidence on each side. The seventy-fifth anniversary debate about the distances to gamma-ray bursts involved a question that nearly everyone thought had already been answered. Lamb convinced no one, and Paczyński did not need to convince anyone.

13

The Theorists' Turn

Like a cigarette smoker or alcoholic who repeatedly postpones the promised day on which he will kick his habit, most astrophysicists had long avoided the problem of explaining how gamma-ray bursts at cosmological distances could work. The only serious attempts, those of Goodman and Paczyński in 1986 (Chapter 10), had predicted properties—a submillisecond pulse of gamma rays narrowly concentrated in a black-body spectrum at a few MeV energy—-very unlike the actual properties observed (Figures 1-2, 10-4). The BATSE results, like an ominous warning from the doctor, forced the theorists to confront reality.

The difficulty was not simply the required energy of 10^{51}–10^{52} ergs. This seems enormous, but is actually less than that released when a neutron star is born, when two neutron stars collide, or when any kind of star is swallowed by a black hole. Each of these events makes available roughly 10% of the rest mass energy (the $E = mc^2$ energy corresponding to the mass m) of the stars. For a neutron star this is about 3×10^{53} ergs, more than sufficient. The difficulty was not even the gigantic required power, probably in the range 10^{51}–10^{52}* ergs per second, although this is nearly a billion times the power radiated by a luminous galaxy. The difficulty, overcome somehow by nature but not yet by the human mind, was to figure out how a small frac-

---

*This is about the same *number* as the total required energy because the peaks of typical bursts last about a second, but power and energy are different physical quantities: power is energy divided by the time it takes the energy to arrive. Power is like an hourly rate of pay; energy like a bank balance. Power may be measured in kilowatts, or ergs per second; energy in kilowatt-hours or ergs.

tion, perhaps 1%, of the energy appears in the form of a gamma-ray burst.

The great energy required by the cosmological distance scale was even a helpful simplification. Any explanation that cannot provide that energy must be ignored. Most of the lush garden of theoretical models that sprang up after the discovery of gamma-ray bursts was mowed down by the blade of the energy requirement.

Theory and observation pointed to two possibilities. One was the collapse of the cores of massive stars. Just how massive they needed to be was in dispute, with estimates ranging from a few times to dozens of times the mass of the Sun. It is known that neutron stars are born in such collapses, as demonstrated by the fact that pulsars are found in the remnants of supernovas, the explosions at the ends of the lives of massive stars. The energy released in collapse is observed in supernova explosions and, in one case (Supernova 1987A in the Large Magellanic Cloud), emerging neutrinos were detected directly. It is likely that other collapsing stellar cores and supernovas make black holes. In fact, no neutron star has been found in many comparatively young supernova remnants, suggesting that black holes may have been born.

The other possible energy source was the collision or coalescence (a gentle collision resulting in merger) of two neutron stars, or of a neutron star and a black hole. In 1975 the first binary pulsar—a pulsar orbiting another star—was discovered. Careful analysis of its orbit showed that its companion was another neutron star. Their orbits are shrinking slowly, as gravitational waves, a form of radiation analogous to electromagnetic radiation but consisting of gravitational rather than electric and magnetic fields, rob them of energy and angular momentum. Gravitational radiation has the same effect on a binary orbit in the vacuum of space that atmospheric friction has on the orbit of an Earth satellite. It was one of the most striking predictions of Einstein's general theory of relativity, the modern theory of gravity, a prediction that was confirmed to better than 1% accuracy by observations of the first binary pulsar. Scores of binary pulsars have been discovered since, although only a few have neutron star companions. The theory works for them also.

As the orbit of a binary neutron star shrinks, it emits more and more gravitational radiation. In a finite and quite calculable time the two neutron stars will draw together until they touch. The first neutron star binary discovered will do this 301,000,000 years from now, but others, formed in the distant past, are even now almost in contact. Emitting a giant burst of gravitational radiation, the two neutron stars will coalesce, making a rapidly rotating flattened superneutron star (super, because it contains the matter of two neutron stars). Enough gravitational radiation will be produced to be detectable by several gravitational-wave observatories now under construction around the world, even if this happens in a distant cluster of galaxies.

The superneutron star is born hot and rapidly rotating, which helps it fight gravity, for it is too massive to settle down as a cold, nonrotating neutron star. Gravitational waves, electromagnetic waves, and neutrinos quickly drain its heat and angular momentum, and its core soon collapses into a black hole. A few percent, or perhaps somewhat more, of the matter is left behind, orbiting the black hole. This matter, drawn from the interiors of the neutron stars, consists mostly of neutrons, and is expected to have a roughly toroidal shape, a dense and massive neutronic doughnut less than 50 km across.

Astronomers who study gravitational waves, as well as those who study gamma-ray bursts, have tried to estimate how often two neutron stars coalesce. The results are very uncertain because they depend on the poorly understood processes that make binary neutron stars. On the basis of the number we observe, a (very controversial) allowance for those too faint or distant to detect, and on how long they have left to live, it is estimated that in a galaxy like ours there is roughly one neutron star coalescence per million years.

How might some of this energy be tapped to make a gamma-ray burst? The first step was taken by Tsvi Piran and his student Amotz Shemi even before BATSE and GRO were launched. Piran had originally been a "relativist," a physicist specializing in general relativity, the theory of gravity. This led him to the study of coalescing neutron stars, in which gravity is strong and the abstruse and technical details of general relativity are essential, and from there to gamma-ray bursts. Like most Israeli men up to the age of 50, he had to take time off each year for military reserve service. His consisted of following a school bus in a jeep to protect it from attack. This duty could be dangerous; at least one Israeli soldier has been killed blocking a car bomb meant for a school bus. It was a reminder to the American visitor that civilization, including scientific research, is only protected by continual vigilance.

When two neutron stars coalesce, an enormous amount of gravitational energy, as much as 10^{54} ergs, about half the $E = mc^2$ energy of the Sun and more than that required to explain gamma-ray bursts even at cosmological distances, may be released. It was unclear just what would happen to this energy (some of it would be carried off by the gravitational waves), but some of it might be available to make a gamma-ray burst. Goodman and Paczyński's calculations of the properties of fireballs produced by the release of pure energy, in the form of gamma rays, into space were the natural place to start thinking about this problem. Shemi and Piran added a little ordinary matter in the form of protons and electrons. This is just like adding hydrogen, the most abundant element in the universe, because a hydrogen atom consists of one proton and one electron. Other elements, with heavier nuclei and more electrons, would have similar effects.

Even a small amount of hydrogen makes a big difference, because while electron–positron pairs completely annihilate into gamma rays, protons remain forever. They cannot annihilate. Further, each proton is accompanied by an electron that balances its electric charge. Now the total number of electrons includes both those created when gamma rays make electron–positron pairs and those that came along with the protons. After all the positrons have completed their suicide pacts with electrons and annihilated, a few electrons will be left over, exactly as many as there are protons. The protons and the surviving electrons form the debris of the fireball.

Without protons, the fireball soup of electron–positron pairs would entirely turn into freely escaping gamma rays. With protons present, the gamma rays are trapped within the cloud of leftover electrons (by Compton scattering; Chapter 5) and cannot escape from the debris. The pressure of the gamma rays pushes on the electrons as the wind pushes on the sails of a ship or the blades of a windmill, and the electrons are accelerated outward.

The fireball debris is a plasma consisting of charged particles, but it has no overall charge because the positively charged protons are balanced by an equal number of negatively charged electrons. If this balance were not exact, the unbalanced charge would produce an electric field pulling in the missing electrons from outside, or expelling any extra electrons, ensuring equal numbers of protons and electrons everywhere. Even an infinitesimal deviation from perfect balance produces a large electric field, and strong forces restoring the balance.

In an internal combustion engine hot gas is trapped by a massive piston and pushes on it. A gamma-ray burst is analogous, with the hot gas made up of electron–positron pairs and gamma rays, and the piston of protons and electrons locked together by their electric charges. Soon the energy of the fireball is transferred to the debris, just as in a gasoline engine the energy of burning fuel is turned into the motion of the pistons, and then into the motion of the entire automobile.

The acceleration in a gamma-ray burst is billions of times greater than in an automobile engine, and the debris is soon accelerated to nearly the speed of light. Just how fast it goes depends on how much debris there is, and how much energy. If there is less debris, it moves faster, while if there is more debris, it moves less rapidly, just as a light racing car accelerates faster than a heavy truck. Typical estimates are speeds between about 99.99% and 99.9999% of the speed of light.

Matter moving this fast has a total energy of between roughly one hundred and one thousand times its rest mass energy of mc^2 (the energy it would have if it were not moving at all). The total energy is usually written γmc^2, where the Greek letter γ (gamma), called the Lorentz factor, is the ratio of the total energy to the rest mass energy. Physicists and astronomers usually think and calculate with the Lorentz factor, rather than the speed. The kinetic energy is the total energy minus the rest mass energy. For Lorentz factors as large as those involved in gamma-ray bursts the rest mass energy is insignificant; nearly all the total energy is kinetic energy.

If matter is added to a fireball, then nearly all of the fireball's energy is converted to kinetic energy of the fireball debris. Almost none emerges as radiation, the astronomical equivalent of heat out the tailpipe. Introducing matter solves one problem—the fact that a naked (without matter) fireball, calculated by Goodman and Paczyński, emits radiation with the wrong properties to explain gamma-ray bursts. It creates another problem—a clothed fireball, calculated by Shemi and Piran, emits almost no radiation at all. The kinetic energy of the debris is, in principle, available to make radiation, but cannot do so if the debris simply coasts through the vacuum of space.

The next steps were taken in 1992, after the first BATSE results were announced, and went in two different directions. The first was followed by Martin Rees in collaboration with Peter Mészáros of Penn State. They offered a path through the roadblock of no radiation at all erected by Shemi and Piran, and also a possible explanation of the most enigmatic property of gamma-ray bursts, their diverse and complex pulse shapes.

The space between the stars is not a perfect vacuum, but is filled with an extremely dilute gas. In our galaxy this interstellar medium has an average density of about one particle (or atom) per cubic centimeter. In some regions it may be a thousand times less dense, and in a few places it is thousands of times denser. All of these are enormously rarefied compared to our atmosphere, in which there are (at sea level) about 3×10^{19} molecules per cubic centimeter. The interstellar medium is a better vacuum than any made in the laboratory. All galaxies must contain some interstellar medium because not all primordial gas is made into stars, and some of it is thrown off again when stars die. Even the space between the galaxies contains some gas, though it must be much more dilute than even the gas within galaxies, and almost nothing is known about it.

The energetic protons and electrons accelerated by the pressure of a fireball will not travel freely forever. They will run into the surrounding interstellar medium. Even though the medium is very dilute, the fireball debris, moving at almost the speed of light, soon sweeps up enough mass to affect its motion.

This "sweeping up" is not a simple matter of collisions like those between billiard balls, or between molecules in the air we breathe. The rates of such collisions are infinitesimal at interstellar densities. If they were the only kinds of collisions, an energetic proton could zip through a galaxy like ours, edgewise, hundreds of times before colliding once.

A single charged particle, or a few of them, or many if they are very dilute and moving in all directions as do cosmic rays, can penetrate deeply into interstellar matter, suffering only these rare collisions between individual particles. A dense plasma, such as fireball debris, cannot. Like a gale whipping the sea into mountainous waves and filling the air with spin drift, when the debris enters another plasma, such as the interstellar medium, it creates giant waves of another kind. These are called plasma waves.

Plasma waves involve a partial separation of the positively charged protons from the negatively charged electrons, and large electric fields pulling them back together. There may also be large magnetic fields. When the fields of many waves are all tangled up and interfere with each other, the result is plasma turbulence. The turbulent fields push on all the charged particles, both those in the fireball debris and those in the interstellar medium.

The two plasmas collide, but the forces between them are carried by turbulent plasma waves rather than by the collisions of individual particles. The plasma collision is a collective process, for the plasma turbulence involves the correlated motion of large numbers of charged particles in each plasma. The particles' individual, one-on-one, interaction is negligible, but collectively they collide as decisively as a dropped egg smashes onto a concrete pavement.

Collisions make heat. The source of this heat is internal friction, occurring when atoms or particles within the colliding substances slide over or around one another. Internal friction is just as real and ubiquitous as the more familiar forms of friction that bring a sliding or rolling object to a halt. If you drop a coin onto the floor, the heating is imperceptible. If you flatten it with a hammer, it feels warm. If you fire a bullet into it, it becomes too hot to touch.

The fireball debris and the interstellar medium collide at the debris's speed, nearly the speed of light. Then three things happen. First, the enormous kinetic energy of the protons in the fireball debris is converted to random motion, in which they move in all directions. This is equivalent to the random motion of the molecules of ordinary air, but in the gamma-ray burst it corresponds to temperatures of roughly 10^{15} degrees! Initially, nearly all this energy resides in the protons, because they are about two thousand times more massive than the electrons. Constituting nearly all the mass of the debris, they carry nearly all its kinetic energy. Second, the protons' energy is shared with the electrons. Because the electrons are so light, this accelerates them much closer to the speed of light than the protons' speed; the electrons' energies are about the same as the protons', but their Lorentz factors are perhaps a thousand times greater, approaching a million. Third, some of the energy is converted to magnetic fields.

Plasma physics is a difficult subject, and plasma turbulence even more so. None of these results is understood in any detail. It is only known that if they are not correct, there will be no observable gamma-ray burst.

Suppose we accept all the assumptions made so far, at least as a working hypothesis. Then a straightforward calculation, using the well-established (and empirically tested) theory of synchrotron radiation (Chapter 5), shows that the burst of energy will make a fireball, which will accelerate high-energy protons, which will collide with interstellar gas, as a result of which energetic protons will give their energy to electrons, which will radiate synchrotron radiation in the form of soft gamma rays with roughly the energy required to make a gamma-ray burst. Each assumption was essential. For example, if the protons' energy were not shared with the electrons the synchrotron radiation would be a million times less powerful, and its frequency a million times lower, infrared light instead of gamma rays. So far, this model consists of a series of straightforward steps building on the results of Shemi and Piran.

Astronomers know that the interstellar medium is far from uniform. Popular astronomy books contain spectacular pictures illustrating a wonderland of wisps, filaments, rings, shells, hourglasses, and loops. Some of these are comparatively dense clouds condensing into new stars. Others are the remains of exploding stars or winds from stars more gently shedding mass. The more dilute general interstellar medium, in which a fireball is most likely to find itself, emits very little light and is essentially invisible. However, radio observations show that it, too, is very clumpy.

Rees and Mészáros's innovation was to suggest an explanation of the complex and diverse time dependence of gamma bursts. They pointed out that when an expanding shell of fireball debris strikes a wisp of denser interstellar matter some of its kinetic energy can be turned into a brief flash of gamma rays. In the deeper vacuum between the wisps little energy would be released and there would be little radiation. This appeared to solve the problem of how a single brief fireball could produce a long gamma-ray burst containing dozens of subpulses. Each subpulse would correspond to a wisp or clump of gas. The diversity of burst time histories was consistent, at least qualitatively, with beliefs about the interstellar medium: in some places homogeneous, in others clumpy, with a very wide range of density.

Vladimir Usov, now at the Weizmann Institute in Israel, went in a different direction. The catastrophic event that produces a gamma-ray burst was generally believed to leave behind a rapidly rotating, probably magnetized, superneutron star or a dense neutronic doughnut orbiting a black hole. In papers published in 1992 and 1994 Usov suggested that such a superneutron star would be a continuing source of energy as its gravity compressed it over a period of seconds, or longer. The same process might also work for the neutronic doughnut, as it was gradually swallowed by the black hole.

The rotating matter was expected to be strongly magnetized because it would have been derived from neutron stars. Pulsars and the (not quite proven) magnetars (Chapter 10) are intensely magnetic neutron stars, and it was reasonable to assume that the neutron stars involved in making gamma-ray bursts would be magnetized too. If some neutron stars were not magnetic, then perhaps they would not make bursts, but others still could.

When a magnetic object rotates it radiates electromagnetic energy at a rate proportional to the square of its magnetic field, the sixth power of its size, and the fourth power of its rate of rotation. This radiation is insignificant for a toy magnet (small, slow, and low-field) but sufficient to slow the rotation of a massive pulsar in thousands or millions of years. The much faster rotation of a superneutron star or a neutronic doughnut around a black hole could radiate all 10^{52}

ergs of its energy in seconds if its magnetic field had the strength suggested for magnetars. The radiation would initially take the form of low frequency radio waves of enormous intensity.

A pulsar radiates steadily because it has a rigid solid crust. Its rotation is regular, and its magnetic field, locked into the crust, is constant. In contrast, a neutronic doughnut or superneutron star is fluid, and its inner parts rotate faster than its outer parts, just as the inner planets revolve about the Sun faster than the outer planets. The magnetic field, tied to the fluid by electric currents within it, is stretched and tangled by the flow, continually changing its strength and form like an unseen Proteus. When the field is large, and suitably oriented, a pulse of electromagnetic energy is radiated. At other times, little is emitted. Impossible to calculate in detail, for it is another form of turbulence, called magnetohydrodynamic turbulence, this was at least consistent with the complex and diverse time structure of gamma-ray bursts.

In Usov's model the mechanism of radiation could be similar to that considered by Rees and Mészáros, for the low-frequency wave would be so intense it would make a series of little fireballs. The fireballs would accelerate debris to relativistic speeds, whose actual values would depend on their power and the amount of ordinary matter injected. When different debris shells moving at different speeds collided with each other, or with the interstellar medium, subpulses of gamma rays would be emitted.

Each of these two models offered a solution to the fundamental problem of gamma-ray bursts at cosmological distances, that of explaining the complex substructure and diversity of the observed bursts with a model based on a single catastrophic stellar death. It was unclear how to tell which was correct, for they both relied heavily on assumptions and rough estimates (often described as "handwaving") about complex and poorly understood plasma physical processes. It was unclear what empirical test could tell which was right.

The answer came from a theoretical insight. Edward Fenimore and co-workers at Los Alamos, and independently Tsvi Piran and his student Re'em Sari in Jerusalem, following ideas that went back to 1974, looked more carefully at the time dependence of the energy released when a relativistic debris shell strikes a clump of interstellar matter. The radiation may rise abruptly, but it will fall slowly (Figure 13-1). This pulse shape is frequently observed, both for gamma-ray bursts as a whole and for subpulses within them.

The reason is shown in Figure 13-2. This is essentially the same as

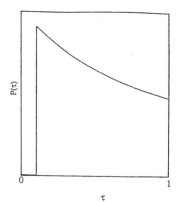

P(τ)

0   1

τ

Figure 13-1. How the power released in a collision between a relativistic debris shell and an interstellar cloud varies with time, as measured by a distant observer. External shock models predict that a burst will have this pulse shape, while internal shock models predict that a subpulse may have this shape. (*Astrophys. J.* V. 422, p. 251 [1994])

Figure 5-2 explaining light echoes, but instead of a point source of radiation there is now a fireball that makes debris traveling at almost the speed of light (rather than light itself). The scattering region is now the interstellar cloud where the debris collides and gamma rays are made (rather than scattering radiation from the point source). Although the physical processes are different, the geometry is the same. An observer sees radiation from a ring that expands with time, starting abruptly as a point (on the horizontal straight line), but only fading gradually.

The fading comes from another consequence of relativity: the radiation produced by particles moving at nearly the speed of light is almost entirely directed in a narrow cone along their direction of motion. At first the observer sees radiation emitted by particles moving exactly toward him. After that passes by, he sees radiation emitted by particles moving slightly sideways, which takes a longer path and

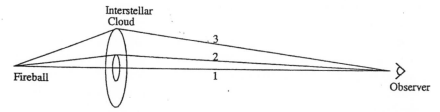

Figure 13-2. The geometry of radiation from a relativistic shock. The debris, moving at nearly the speed of light, is produced at the fireball and collides with an interstellar cloud or clump. The observer first sees radiation that has traveled along the straight line path 1, then in an expanding ring, following paths (2, 3) that progressively deviate more and more from the straight path, arriving later and later.

arrives later. Once the ring has expanded some more, it sends him very little light (because he is no longer close to the direction of motion of matter in the larger ring), and he observes the burst to dim and disappear.

This has yet another important consequence. An observer sees only radiation emitted by matter moving almost exactly towards him. He cannot tell if the relativistic debris forms a complete shell, moving in all directions, or only a small piece of a shell moving toward him, just as you can tell if someone shines a light on you but not if he also shines light away from you up into the night sky. If only small pieces of shells are produced, the total power and energy of the gamma-ray burst are much less than if the shells are complete. However, the bursts certainly don't choose us as targets for their radiation; if the shells are incomplete, there must be many more bursts than we observe, most of them directing their radiation in directions that miss us.

If there were only one cloud in the path of the debris shell (or piece of shell, but astronomers usually simply refer to a shell), the burst would consist of a single pulse, rapidly rising and slowly falling. Some bursts are like this. However, it is not possible to explain the spiky multipeaked gamma-ray bursts unless the clouds are very small and widely separated, so that the individual subpulses are brief and separated by long intervals in which little energy is released. The clouds must be so sparse that most of the debris will never strike a cloud at all. This would be terribly inefficient, making bursts too faint to observe at cosmological distances. The only possible conclusion was that the time dependence of multipeaked bursts cannot come from collisions with a clumpy interstellar medium. Rather, it must represent variations in the rate at which some central engine, such as Usov's hypothesized superneutron star or a neutronic doughnut, emits energy.

Astronomers were soon distinguishing among different kinds of shock models. External shocks result when a relativistic debris shell strikes interstellar matter. Internal shocks result when different debris shells, produced by a central engine of fluctuating power, strike each other.

In each case the collision is assumed to produce a shock, which physicists define as an abrupt jump in the density, temperature, pressure, velocity, and other physical properties of a substance. Some shocks are strong. Other shocks are weak. For example, the sonic boom produced by a supersonic airplane is a shock in the air, but in the typical sonic boom the pressure and other physical properties change by less than one part in a million (near the airplane which makes the boom the changes are much greater). On the other hand, next to a piece of detonating explosive the shock is very strong and the changes are enormous and destructive.

We do not know in any detail what happens when two plasmas collide at relativistic speed. Calling it plasma turbulence is a confession of defeat, not a statement of understanding. Astrophysicists generally assume a shock is formed. This assumption is made because it is an enormous simplification; it permits replacing the problem of plasma turbulence by comparatively simple statements that on each side of the shock the plasma may be completely described by a few simple and familiar physical variables: temperature, pressure, density, and magnetic field. Plasma turbulence is incalculable, but physicists have studied and understood shocks for more than a century.

An external shock model would predict that toward the end of a burst the Lorentz factor (equivalently, the energy) of the debris shell should gradually decrease as it sweeps up more and more interstellar matter. The burst's spectrum should soften (shift to lower-energy gamma rays) and its subpulses should be stretched out. An internal shock model predicts none of these effects. Empirically, in many bursts the spectra do soften with time, but the subpulses, when present, remain sharp and narrow. Neither simple external shock models nor simple internal shock models are sufficient.

Pricked by the spurs of observational data, theorists galloped off to develop internal and external shock models of ever-increasing complexity. These models all had in common matter moving at relativistic speeds, with Lorentz factors of hundreds. This was a novel idea in astrophysics. Individual relativistic particles with enormous Lorentz factors had been known for a long time, in the cosmic rays (discovered in 1912) and as the source of astronomical synchrotron emission (discovered in the 1940s), but bulk matter moving so fast was unprecedented. The models differed in the question of what was colliding in the shocks. In internal shock models it was two separate shells of relativistic debris. In external shock models it was a shell of relativistic debris and the interstellar gas, initially at rest.

Relativistic bulk motion was necessary for two reasons. The first was that a relativistic shock appeared to be the only process that could make the observed bursts as synchrotron radiation, and no other process offered any hope at all. The second was to avoid the compactness problem (Chapter 5). Only if the radiating matter were moving almost exactly outward, at relativistic speed, would its synchrotron gamma rays be narrowly enough beamed, and dilute enough (produced far enough from the central engine) to avoid destroying each other by making electron–positron pairs. These two arguments were mutually consistent, and even gave similar estimates of the speed and energy of relativistic motion, with a Lorentz factor in the range 100–1000.

Now it was time to test the models.

14

Afterglows

By 1993 most astronomers were convinced, on the basis of the BATSE data and apparently sound theoretical arguments, that gamma-ray bursts occur at cosmological distances, and that they involve relativistic expansion with Lorentz factors of one hundred or more. A convincing argument is only the beginning; it needs empirical confirmation.

Cosmological distances had a straightforward, and testable, implication—that if lines could be found in the spectra of gamma-ray bursts, they would show large redshifts. Their wavelengths would be multiplied by factors of 1.5, 2, 3 or more compared to their laboratory values. However, the spectra of bursts appeared (aside from the controversial, and unreplicated, reports of cyclotron and annihilation lines; Chapter 7) to have no lines at all. This was consistent with the popular theory that the observed emission is the result of synchrotron radiation, which has a smooth lineless spectrum, but did not provide any confirmation (or disproof) of either cosmological distances or relativistic expansion within the bursts themselves.

For a spectral line to demonstrate a cosmological redshift, it would be necessary not only that its redshifted wavelength be measured, but also that its wavelength at rest (with no Doppler shift) be known. This would generally not be the case for cyclotron lines, because they could have any rest wavelengths. To determine their rest wavelengths independently would require independent quantitative knowledge of the strength of the magnetic field, which does not exist.

Optical astronomers know that atoms and ions produce numerous spectral lines with characteristic *ratios* of their wavelengths. These ratios are not changed when the wavelengths are redshifted, so observation of a characteristic ratio (e.g., 27:32 in hydrogen or 1:1.002576 in ionized magnesium) immediately tells the observer which lines he is observing and their rest wavelengths. Simply dividing the rest wavelengths into the observed wavelengths and subtracting one yields the redshift.

In principle, this method could be applied to nuclear gamma-ray lines, which also have characteristic wavelength ratios, but none of these was ever observed in a gamma-ray burst. The positron annihilation line might also be used. Because it is the product of a basic physical process likely to occur anywhere there is enough energy to make gamma-rays, no sophisticated comparison of the wavelengths of multiple spectral lines would be required to identify it. The reported annihilation lines (Chapter 7), if real, appeared to imply little or no cosmological redshift, but they were never confirmed and were generally disbelieved. As a result, they were not considered strong evidence against cosmological distances, but they certainly could not be used in their favor either.

The astronomer's usual method of solving problems like this is to find visible counterparts to the bursts. Almost every astronomical object observed in visible light shows numerous spectral lines, and this is the way quasars were discovered. If gamma-ray bursts were located in distant galaxies the redshifts of the galaxies could be measured using methods of visible spectroscopy honed for a century. Twenty years of effort had so far failed to determine even a single burst's position accurately enough to identify it with one of the enormous number of distant galaxies that carpet the sky.

Testing the hypothesis of relativistic expansion would not directly confirm or refute the hypothesis of cosmological distances, but would at least establish if the theorists were on the right track. The expansion itself might be expected to produce an enormous blueshift (lines would be observed at wavelengths shorter than those measured in the laboratory), but the theory of synchrotron radiation and the data concurred that there were no lines to measure.

However, the theory did make definite predictions about the shape of the distribution of energy with frequency, known as the continuum spectrum. In fact, these predictions were different for gamma-ray bursts, because of their relativistic motion and relativistic shocks, than for other astronomical sources of synchrotron radiation. The difference arose from the nature of the process that accelerates the electrons that make the synchrotron radiation.

Astronomers were familiar with the problem of cosmic-ray acceleration, which had been an enigma for many years. These extremely energetic particles, mostly protons but also a few heavier nuclei and electrons (with Lorentz factors up to 10^{11}), fill interstellar space. How they were accelerated was a complete mystery for decades after their discovery in 1912, until in 1949 Enrico Fermi offered a partial solution. He showed that if fast particles were scattered by (bounced off) slowly moving interstellar clouds the particles would, on average, gain energy. The actual force between the particles and the clouds would be carried by magnetic fields.

The energy required to accelerate the particles would be drawn from the kinetic energy of the clouds' motion. In an individual scattering (bounce) a particle might gain (or lose) only 0.01% of its energy, and the gains would outnumber the losses by about 0.01%. It would take about a billion such bounces for a particle to double its energy. However, if there were enough (billions of) bounces, a particle's energy could be multiplied until it reached the enormous values observed in the cosmic rays.

In Fermi acceleration the number of particles accelerated to an energy E is proportional to a power of E. This is naturally called a power law distribution. Usually the power is a negative number between -2 (E^{-2} is the reciprocal of $E \times E$) and -3 (E^{-3} is the reciprocal of $E \times E \times E$), meaning that only a very few particles are accelerated to high energies. For example, if a certain number of particles are accelerated to an energy E, then fewer than one-fourth as many are accelerated to the energy $2E$, and fewer than one-sixteenth that many to the energy $4E$, and so on. The distribution of energies of the cosmic rays is observed to follow a power law, a fact that was mysterious until Fermi explained it. The spectrum of synchrotron radiation in many astronomical objects implies that the electrons that make it also have a power law distribution of energies.

Power law distributions are very common in nature, and are found whenever there is no special value of the energy (or whatever is being measured) setting a natural scale. For example, the distribution of the sizes of earthquakes is a power law known as the Gutenberg–Richter law, and the distribution of incomes in a society also follows a power law, at least for the middle and upper classes. Welfare and the minimum wage set a natural scale (a floor) for low incomes, so these do not follow the power law that describes the distribution of higher incomes.

By the late 1970s it was realized that the interstellar clouds whose motion accelerates the cosmic rays were, in fact, probably the material on each side of an interstellar shock produced when supernova debris runs into the interstellar medium. The rate of acceleration is then much higher, because if a charged particle were magnetically trapped at the shock it would gain energy in every bounce. This was called nonrelativistic shock acceleration (nonrelativistic because supernova debris moves at speeds very much less than the speed of light), but a modified form of Fermi's theory still applied.

In nonrelativistic shock acceleration only a tiny fraction of the charged particles present are accelerated at all. For a typical supernova remnant shock this fraction might be one in a million. These few particles are accelerated to relativistic speeds and large Lorentz factors, but there is not enough energy to accelerate all the particles, so the rest remain cool. "Cool" may actually be quite hot by ordinary standards (10^8 degrees in a supernova remnant), but their Lorentz factors are stuck in the range between 1 (at rest, with no kinetic energy at all) and 1.01 for electrons and between 1 and 1.00001 for protons.

Because fireball debris is moving at nearly the speed of light it makes relativistic shocks when it collides with the interstellar gas. In such a shock nearly all the particles are accelerated to Lorentz factors much greater than one. The details are not understood, but the acceleration is believed to occur in a single step or a few large steps, rather than in an enormous number of tiny steps slowly pushing a few particles up the energy ladder. In fact, the Lorentz factors of the individual protons accelerated by a relativistic shock approximate that of the fireball debris itself, typically in the range 100–1000, as the Lorentz factors of the electrons are about a thousand times greater. There will be comparatively *few* particles at lower energies, in contrast to Fermi acceleration in a nonrelativistic shock, in which the less energetic particles vastly outnumber the more energetic ones.

These differences between nonrelativistic and relativistic shocks affect the synchrotron radiation of the electrons they accelerate. The spectrum of synchrotron radiation produced by a nonrelativistic shock is itself a power law. The intensity of radiation at each frequency is proportional to the $-s$ power of the frequency (the minus sign is part of the customary definition of s). This is the same mathematical function used in Fermi acceleration, but now it is the distribution of radiation that is described by a power law, while in Fermi acceleration it was the distribution of particle energies.

This type of spectrum is observed in many astronomical objects, and is usually the strongest argument that they are emitting synchrotron radiation. The value of s describes how much radiation there is at high frequencies compared to that at low frequencies; large s means that there is more low-frequency radiation, and small (or even negative) s means that there is more high-frequency radiation. Usually, s

is near $+0.6$, and it is generally in the range $+0.5$ to $+1.0$. Such a spectrum contains a comparatively large fraction of its energy in lower-frequency radiation, a consequence of the fact that nonrelativistic shocks accelerate many more lower-energy particles (which radiate lower-energy photons) than extremely high-energy ones.

In contrast, the synchrotron radiation from a relativistic shock is predicted to have a single special spectrum at lower frequencies. That spectrum is also a power law, but with $s = -\frac{1}{3}$. It does not matter whether the shocks are internal or external. The reason is that there are too few of the intermediate energy electrons to be significant. Even at lower frequencies, the source of the radiation is still the highest-energy electrons, which radiate this almost universal spectrum. This spectrum is observed in laboratory particle accelerators, in which all the electrons have the same high energy.

For the first time a model of gamma-ray bursts made a testable quantitative prediction about their properties. The $s = -\frac{1}{3}$ spectrum predicted by relativistic shock models extends from soft gamma rays through X rays, visible and infrared light down to radio frequencies. Simple extrapolation from the gamma-ray intensity led to a prediction of the intensity of bursts in visible light and at radio frequencies. A bright burst might have a magnitude in visible light in the range 16–18 during the burst itself. This would be easily detectable by a small professional telescope (or even a large amateur one), if only the astronomer knew where and when to look. But, of course, he did not. It was much too faint to be recorded in most archival sky photographs.

At higher frequencies (harder gamma rays) the $s = -\frac{1}{3}$ spectrum was not expected to hold. Electrons radiating at these frequencies lose their energy quickly, which increases the number of electrons of intermediate energies and the spectrum they radiate. The predicted value of s is then $\frac{1}{2}$. A few electrons are accelerated to sufficient energy to radiate still-higher-frequency radiation, so the spectrum there may drop off sharply.

As long as gamma-ray bursts were observed only in soft gamma rays these predictions of relativistic shock models could only be tested at those energies. The first attempts were made using gamma-ray data from BATSE. Some bursts showed a clear $s = -\frac{1}{3}$ spectrum at the lowest gamma-ray energies, changing to $s = \frac{1}{2}$ at intermediate energies, and dropping rapidly at the highest energies, just as predicted. Others were close to $s = \frac{1}{2}$ at all observed energies, or lay between $s = -\frac{1}{3}$ and $s = \frac{1}{2}$, behavior consistent with the predictions.

This evidence was encouraging, but not compelling, and may have been affected by the uncertain procedures used to deconvolve the BATSE spectra. A different analysis, fitting spectral curves to the data, found evidence for s less than $-\frac{1}{3}$ in a few bursts, apparently inconsistent with synchrotron emission. This was hard to understand in any model, but cast further doubt on the entire uncertain process (Chapter 7) of extracting reliable spectral information from NaI(Tl) detectors, such as BATSE's.

In a fireball model the debris will gradually sweep up interstellar matter, as a broom or mop accumulates an increasing load of dirt. The Lorentz factor of the debris will decrease as it slows. In an external shock model this is how the gamma-ray burst is made; in an internal shock model it occurs after the burst itself; in either case it is inevitable.

As the Lorentz factor decreases so do the energies of the accelerated particles and the frequency of their synchrotron radiation. It will gradually slide from gamma rays to X rays of progressively decreasing frequency, ultraviolet light, visible light, and then to infrared light. This is shown in Figure 14-1. As its frequency decreases the radiation lasts longer and longer. Finally, after the fireball debris slows to nonrelativistic speed the synchrotron radiation peaks at radio frequencies. This final stage of a gamma-ray burst resembles a supernova remnant, in which a nonrelativistic debris shell sweeps up interstellar matter.

By 1994 it was apparent that relativistic shock models of gamma-ray bursts thus made a series of remarkable predictions. Following the burst proper there should be continuing emission at progressively lower frequencies (in external shock models the burst itself is only the initial stage of this progression). At all times the instantaneous spectrum should have the special $s = -\frac{1}{3}$ form at frequencies below the peak of emission, extending down to radio frequencies. Finally, it was also possible to predict that at lower radio frequencies the intensity should drop off steeply (with $s = -2$) because the very electrons that emit synchrotron radiation also absorb it, a phenomenon known as self-absorption. This is observed in many astronomical synchrotron sources, but with $s = -2.5$; again, the special nature of relativistic shock acceleration predicts a different spectrum.

For the first time, theories of gamma-ray bursts were able to make specific testable predictions, rather than simply offering speculative explanations of what had already been observed. Predictions must be tested by observation. Unfortunately, the afterglow (as it came to be called) of gamma-ray bursts was predicted to be faint and short-lived, and the BATSE coordinates were much too approximate to permit

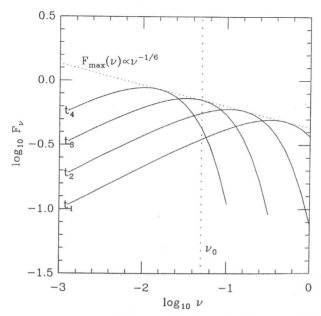

Figure 14-1. Theoretically predicted evolution of an afterglow spectrum. The units are arbitrary. At first (time t_1) the spectrum peaks at high photon frequencies (to the right). At later times (t_2, t_3, t_4 etc.) the spectral peak moves to lower frequencies (to the left). At any fixed frequency ν_0 the intensity first rises, then peaks (at t_3 in the figure), and finally falls. This behavior has been observed, at least qualitatively, although most afterglows are discovered after the spectral peak has passed through the visible region to the infrared, and only the decay is seen. (Reprinted by permission from Fourth Huntsville Gamma-Ray Burst Symposium p. 694 © 1998 American Institute of Physics.)

visible or radio telescopes to find them on the sky, among millions of unrelated astronomical objects.

Progress came in 1996 with the launch of an Italian–Dutch satellite called BeppoSAX. It was designed chiefly to study X-ray astronomy, not gamma-ray bursts, but it turned out to be uniquely well suited to discovering afterglows. It succeeded because it included three very different kinds of instruments that were extremely powerful when working together.

BeppoSAX carried two wide-field cameras, each of which would detect X rays and soft gamma rays from a broad area of sky about 40° across. That is about the apparent size of a beach ball held at arm's length, and covers about 2% of the sky. There were also several narrow-field instruments (sensitive to photons of different energies) that could study small regions about a degree across, twice the diameter of the Sun or the Moon. Finally, there was a gamma-ray-burst monitor. This was actually an auxiliary component of one of the narrow-field instruments, designed to tell when charged particles from the Earth's Van Allen radiation belts were making spurious signals in the X-ray detector, but it would also detect gamma-ray bursts from any direction.

BeppoSAX succeeded because these instruments worked together. The gamma-ray-burst monitor alerted the ground controllers to the presence of a burst. Usually, it was not detected by any other instrument on board because none would be pointing in its direction. However, roughly one burst in twenty occurred within the field of view of one of the two wide-field cameras.

The most important property of these instruments was that they determined the direction of the arriving X rays (the low end of the soft gamma radiation emitted by a gamma-ray burst). This radiation cannot be focused, so a technique long known to X-ray astronomers was used. A perforated mask is made of a sheet of metal that absorbs X rays. Roughly a third of its area is removed to make a pattern of holes in apparently random but carefully measured locations. This mask is mounted in front of the detector, and its solid parts cast shadows when X rays pass through its holes. From the distribution of the shadows on the detector the direction to the source of radiation can be calculated, just as you can use the length and direction of shadows on the ground to determine where the Sun is in the sky without looking at it. If a gamma-ray burst were detected, its position might be measured to within a circle 6 minutes of arc in diameter, a fifth the apparent size of the Sun and Moon and about the same as a dime 30 feet away.

This combination of wide field of view and accurate measurement of position was crucial. If the field of view had been smaller, hardly any gamma-ray bursts would have been detected; had it been larger, the positions of the bursts could not have been measured nearly so accurately.

Once the wide-field camera had located a burst, the narrow-field instruments could be pointed in its direction. This required commands from the ground, and took several hours. The narrow-field instruments could then locate the burst (by that time, probably its X-ray afterglow if there were one) to an accuracy of about 1 minute of arc, smaller than the apparent size of Jupiter. This accuracy would be unprecedented for a gamma-ray burst. More important, the coordinates of the burst could be transmitted to observers on the ground within hours.

All the pieces came together on February 28, 1997. Within 21 hours of a burst a large international team (thirty-one scientists representing eight countries were credited) was able to locate it on the sky and point large telescopes at it. This team was led by the Dutch astronomer Jan van Paradijs, who died two years later at the age of 53.

He had previously been best known for the discovery of the optical emission of X-ray bursts, more than two decades earlier. X-ray bursts occur on accreting neutron stars in our galaxy as the result either of runaway nuclear reactions or of a sudden increase in the accretion rate. Their optical emission is the result of X-ray energy absorbed by a binary companion star (the source of matter for accretion), and re-radiated as visible light. This is the same mechanism assumed to make optical counterparts of gamma-ray bursts when they were believed to occur on nearby galactic neutron stars (Chapter 8). X-ray bursts are now known to be unrelated to gamma-ray bursts despite their similar names and spectral regions of observation, although as recently as 1990 a sophisticated theoretical attempt was made to explain gamma-ray bursts as similar events.

Van Paradijs's team discovered the faint (then 21st magnitude, about a million times dimmer than the faintest star visible with the naked eye) visible counterpart of a burst. An image obtained by the Hubble Space Telescope about a month afterward (when it had faded to 26th magnitude, yet another hundred times fainter) is shown on the dust jacket of this book. The nearby fuzzy patch is probably the galaxy in which the burst occurred. The gamma-ray burst itself had lasted only about a minute. The narrow-field instruments and, most dramatically, visible-light telescopes, had discovered the afterglow predicted three years before.

The first time is always the hardest. BeppoSAX continued to detect and locate gamma-ray bursts, and optical astronomers discovered that many, but not all, of them have observable afterglows. Beginning with a burst on May 8, 1997, radio astronomers also succeeded in detecting afterglows. It took several days after the burst for the afterglows to become bright enough to observe at radio frequencies, probably as a consequence of the predicted self-absorption while the expanding fireball debris cloud was still small and dense. The intensity of the radio signal fluctuated violently from day to day (radio astronomers can observe in daylight as well as at night, because the Sun is a comparatively weak radio source), a result of bending of the radio waves by the interstellar plasma in our own galaxy, which is analogous to the twinkling of starlight as it passes through Earth's atmosphere.

Once afterglows were detected from visible light to radio waves it became possible to test one of the quantitative predictions of the relativistic shock models, the $s = -\frac{1}{3}$ spectrum. The first measurements, on the afterglow of a burst on May 8, 1997, the second afterglow to be discovered, gave the result $s = -0.44 \pm 0.07$, meaning that the value of s was probably in the range from -0.51 to -0.37. A range is given because all measurements have some uncertainty (in this case systematic rather than statistical), whose size is estimated (with some uncertainty itself!) from knowledge of the properties of the instrument. The predicted value was a little outside the estimated uncertainty range of the measurement, but not very much. The theory was not exact either. Independent measurements of the same afterglow yielded $s = -0.25 \pm 0.04$. Again, this result was close to the prediction, but differed by a little more than the estimated uncertainty. Averaging the two results yielded $s = -0.345$, in excellent agreement with the predicted $-\frac{1}{3}$. The relativistic shock model was confirmed.

Figure 14-2 shows the spectrum, from radio to X rays, of this afterglow. At the lowest radio frequency no signal is seen, probably because of self-absorption. At higher radio frequencies $s = -0.44 \pm 0.07$ is observed. No signal is detected over a broad range between radio waves and infrared light in which instruments are less sensitive (much of this range is blocked by absorption in Earth's atmosphere). A persistent afterglow, dropping more steeply with increasing frequency, is observed in infrared light, visible light, and X rays.

Visible afterglows faded to invisibility in days, as the peak of their radiation moved through the visible spectrum ever farther into the infrared, as expected (Figure 14-1). Close inspection of their positions often showed small, faint, unvarying fuzzy patches of light left behind. These looked like galaxies at cosmological distances, but the appearance of a small faint fuzzy patch of light on the sky does not, by itself, prove anything.

Astronomers demanded proof, and soon found it. On May 11, 1997, observers using the 33-foot (10-m) diameter Keck telescope (along with its twin, the world's largest) obtained the visible spectrum of the afterglow of the burst of May 8, 1997. They found absorption lines with the characteristic wavelength ratios of the lines of iron and magnesium, typical of interstellar matter. The lines were not at the wavelengths measured in the laboratory, but at wavelengths 1.835 times greater; in other words, at a redshift of 0.835. Light from the gamma-ray bursts had been absorbed by matter nearly halfway across the universe.

The absorbing matter might be in the same galaxy as the gamma-

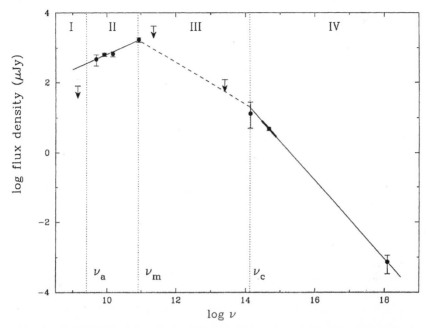

Figure 14-2. The spectrum of the afterglow of the burst of May 8, 1997, observed 12 days later. The theory (solid and dashed lines) is confirmed by the data (solid dots are actual measurements; downward-pointing arrows are measured upper limits). This is a "log-log" plot, in which a power law appears as a straight line. At the lowest radio frequencies (region I) the signal is too weak to detect, falling below the straight-line extrapolation from region II, confirming the predicted self-absorption (ν_a is the frequency below which self-absorption is important). At higher radio frequencies (region II) a power law with $s = -0.44$ ± 0.07 is observed, close to the predicted $s = -\frac{1}{3}$. The spectrum peaks near the frequency ν_m. In region III detectors are insensitive and there are only upper bounds to the spectrum; the dashed line assumes the predicted $s = \frac{1}{2}$, and is consistent with the data. There is a steeper drop off in region IV, from infrared frequencies (ν_c) through visible light (the heavy slanted line segment, representing several observations) to X rays, where a decaying afterglow persists. (Reprinted by permission from *Astrophys. J. Lett.* V. 500, p. L98 © 1998 American Astronomical Society.)

ray burst, in which case the burst also had a redshift of 0.835. Or, the light of the afterglow may have accidentally passed through the absorbing matter on its way to Earth. In that case the burst was even more distant and had a greater (though unmeasured) redshift. In either case, there no longer could be any doubt that gamma-ray bursts occur at cosmological distances. That question was answered conclusively.

Even the handful of diehard supporters of galactic distances retreated to the rather desperate position that while some bursts (all for which there was evidence) were at cosmological distances, some of the others might be galactic. They might even be a variety of soft gamma repeaters. This was impossible to disprove, for there would always be bursts (in fact, the vast majority) for which there was no direct evidence establishing their distances. However, most people only need to be burned once to conclude that all fires are hot.

That same year a much more distant burst was found. The afterglow of a gamma-ray burst on December 14, 1997, faded in a few days, leaving behind a steady patch of light between 25th and 26th magnitude, one of the faintest objects ever to have its spectrum measured. The observations, again using the giant Keck telescope, required fifteen separate half-hour exposures. Combining them, the data showed the spectrum of a galaxy at a redshift of 3.42. This confirmed that the faint fuzzy patch was actually a galaxy. Its redshift was almost certainly the actual redshift of the burst, rather than a lower bound.

As I write this chapter, the record for the redshift of a gamma-ray-burst afterglow is 4.50, held by the afterglow of a burst observed January 31, 2000. This is less than the record redshift of a galaxy or quasar, which is around 6 (but, like most athletic records, frequently improved on by small amounts), but is in the same ballpark. Thousands of galaxy and quasar redshifts have been measured, but comparatively few redshifts of gamma-ray burst afterglows, so it is not surprising that the afterglows do not hold the record. Nevertheless, it is clear that gamma-ray bursts come from remote parts of the universe, and that they happened when the universe was only a fraction of its present age.

Such large redshifts and enormous distances imply that the gamma-ray bursts were extremely luminous and energetic. Although distant, the burst of December 14, 1997, was fairly bright. A straightforward calculation led to the result that it had emitted, chiefly in soft gamma rays, 3×10^{53} ergs, roughly a hundred times greater than the usual estimates of gamma-ray burst energies in the 10^{51}–10^{52} erg range. This would require the accretion onto a black hole of nearly an entire solar mass, or the complete conversion of nearly a tenth that much mass to energy, which strained credulity.

The obvious explanation was that the radiation was narrowly beamed in our direction, reducing its total energy requirement. This also implied that for every burst we observe there must be many we do not observe, because they are beamed in other directions. Beaming is quite possible in a fireball produced by a neutronic doughnut, for the central hole may form a narrow throat, channeling the fireball debris like the nozzle on a garden hose. Similar beaming is observed in quasars and other objects powered by accretion onto a black hole, where it makes long slender jets observed by radio telescopes and even in visible photographs (as long ago as 1918 in the galaxy M87).

Beaming complicated the task of the theorist. Now he had to consider not only the energy and time dependence of the fireball debris, but also its degree of beaming. Calculations that were tractable when spherical symmetry was assumed now became very difficult. It also meant that any results would depend on many assumptions, including the properties of the nozzle that channeled the debris; attempts to compare the theoretically predicted properties of gamma-ray bursts, and especially of afterglows, to the data might degenerate into parameter fitting rather than being clear tests of theoretical ideas.

15

A Supernova Connection?

By the spring of 1998 most of the pieces of the gamma-ray-burst puzzle seemed to be in place: Bursts are produced when dense matter, a neutronic doughnut, is accreted by a black hole. This doughnut makes a series of pair fireballs that accelerate thin shells (or portions of shells) of matter to high energy, with speeds closely approaching the speed of light. This matter is probably directed into jets, but only roughly, like water from a fire hose rather than light from a laser. The gamma rays themselves are made when these shells collide with each other, the faster overtaking the slower, or with surrounding gas. An afterglow is produced as the shells, sweeping up this gas, gradually slow. This outline appeared clear, but the physical details depended on plasma and magnetohydrodynamic turbulence, which are not understood.

It had taken 25 years for the understanding of gamma-ray bursts to reach the level of agreement over fundamentals that the understanding of quasars had reached in 1964, a year after their discovery, when Edwin Salpeter argued that they were the consequence of accretion onto supermassive black holes at the centers of galaxies. Gamma-ray-bursts appeared to resemble quasars, concentrated into smaller sizes and masses and bursting with vastly greater power. Most discouraging for the gamma-ray-burst astronomer was the fact that after a third of a century of hard work the basic physical processes in quasars remained mysterious, or at least controversial. For example, it was still not known how they accelerate energetic particles. Progress in understanding gamma-ray bursts was likely to be even slower because of their unpredictability. Once you have discovered a quasar it will always be there, in the same place, waiting for you to collect more data.

A number of important astronomical questions about gamma-ray bursts remained, quite apart from the mysteries of turbulence, and perhaps easier to answer. One was the origin of the neutronic dough-nut. Most astronomers assumed it was the coalescence of two neutron stars or the swallowing of a neutron star by a black hole. However, in 1993 Stanford Woosley, of the Lick Observatory in California, sug-gested that it could be the collapse of the core of a very massive star, one that was several tens of times the mass of the Sun. Such a col-lapse, accompanied by the explosion of its outer layers, might be an extra powerful supernova. Bohdan Paczyński later called this hypo-thetical event a hypernova, hijacking a term that other astronomers had occasionally used for at least three different kinds of events, all presumably unrelated to gamma-ray bursts.

This would have remained a parochial disagreement among the theorists but for the intervention of fate. On April 25, 1998, a wide field camera on BeppoSAX located a gamma-ray burst (named GRB980425 after its date of detection, as is customary). Its position was quickly transmitted to optical and radio astronomers, who searched for an afterglow. The optical astronomers came up with a big surprise. Right in the circle of possible positions of the gamma-ray burst was an unusual object. It was not the very faint afterglow (typically about 20th magnitude) astronomers had come to expect. Instead, it was 14th magnitude when observed, hundreds of times brighter. Measurement of its spectrum and of its varying intensity immediately revealed it to be a supernova, named SN1998bw (SN means supernova, and bw is essentially a serial number, labeling it the 75th supernova catalogued in 1998).

SN1998bw had a measurable cosmological redshift, but it was very small, only 0.0085, implying that its distance was only about 1% of the other measured distances of gamma-ray bursts. This was consis-tent with its brightness (supernovas at redshifts around 1 are usually about 24th magnitude), but was certainly not expected, based on ex-perience with afterglows, for a gamma-ray burst. If the burst really had the same redshift as the supernova, it was apparently thousands of times less luminous than other gamma-ray bursts, for even though extraordinarily close it was not extraordinarily bright. Further, it was clear that the supernova was not an afterglow. The spectra of after-glows were completely different from that of SN1998bw, and after-glows faded in days, in contrast to SN1998bw, which brightened for about 2 weeks and then faded slowly, reddening and losing about a magnitude a month.

Were the gamma-ray burst and the supernova really the same object, or had they simply appeared in roughly the same part of the sky, at about the same time, entirely by accident? This question was impossible to answer with confidence. The data are shown in Figure 15-1. The wide-field camera position of this burst was rather approximate, a circle 16 arc minutes in diameter, more than twice as large (and with more than four times the area) of some of the more accurate positions measured by that instrument. The narrow-field instruments found two X-ray sources, one steady and one fading, inside the circle of possible gamma-ray burst positions. Radio telescopes found five radio sources (including the supernova). Visible light astronomers found the supernova. That is a total of seven distinct objects. At most one of them could be the counterpart of the gamma-ray burst. Perhaps none were.

Gamma-ray bursts are rare, and so are supernovas. When two rare

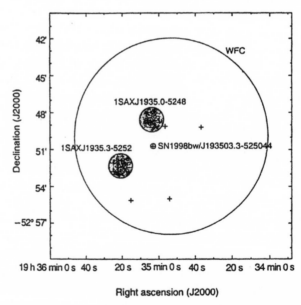

Figure 15-1. The sky around GRB980425. The large circle is the error circle of the BeppoSAX wide-field camera; the gamma-ray burst was almost certainly inside this circle, although this is not quantified statistically. The position of the supernova SN1998bw is marked by the tiny circle with the cross inside. The shaded circles are the error circles of two X-ray sources (one fading as an afterglow might, one steady) detected by the BeppoSAX narrow-field instruments and the crosses are the positions of radio sources. At most one of these radio, visible, and X-ray objects can be associated with the gamma-ray burst, but which one? (Reprinted by permission from *Nature* V.395, p. 665 © 1998 Macmillan Magazines Ltd.)

objects are found together it is natural to conclude this is no accident and that they are related. This was the argument that led to the identification of the burst (now classed as a soft gamma repeater, but nonetheless a rare event) of March 5, 1979, with a supernova remnant in the Large Magellanic Cloud (Chapter 6). That identification was originally based on the unlikelihood that their coincidence in space was accidental, and is now generally accepted as correct. The position of GRB980425 was much less certain—its error circle had about one hundred times the area of the error box of the March 5, 1979, event (Figure 6-1)—but supernovas are so unusual that finding one, even in an error circle of this size, is unlikely to be an accident. Allowing for the fact that the burst and the supernova occurred, as well as could be determined (to a few days' accuracy) simultaneously, led to the estimate that this had no more than one chance in ten thousand of happening accidentally.

This is called an a posteriori statistical argument, and is very risky. The problem (just as for the claims of anisotropy discussed in Chapter 12) is that the statistical test was defined *after* the observations are made, and with knowledge of their results. Many possible hypotheses and tests were implicitly considered and discarded, and only the one that showed apparently extraordinary results was kept. Hence it is not nearly so significant statistically as it might appear.

For example, I take a dollar bill from my pocket. Its serial number is 46176605. There are 100,000,000 possible serial numbers, all equally likely,* so the chance that I will find the number to be 46176605 is only one in 100,000,000. Amazing! Well, it would have been amazing if I had guessed the serial number before looking at it, but I did not. The dollar bill had to have *some* serial number (finding one without a serial number would be a genuinely amazing printer's error). Choosing the number a posteriori, after I have looked at it, I am certain (100,000,000 chances out of 100,000,000) to get a number with only one chance in 100,000,000 of turning up, because any single number has only that tiny probability. I could also try to argue that 46176605 is a very special number (the digit 6 appears three

---

*This is not exactly true, because in U.S. currency several hundred million bills (properly, Federal Reserve Notes) may be printed, reusing the eight-digit serial numbers (a letter code keeps track of the repetitions). The smaller numbers may be used once more than the larger numbers if the total number printed is not an exact multiple of 100,000,000. In addition, different batches are distributed in different parts of the country. These complications do not affect the essential point.

times, which some people think is a sign of Satan . . .) but with a little time and imagination I can concoct an argument that any eight-digit number is special.

If astronomers had found a pulsar, the planet Jupiter, a quasar, a variable star (of any of dozens of varieties), or any other object (and almost any astronomical object is "unusual" or "peculiar" if studied closely enough, simply because, like people, each is a little different from its fellows) near a gamma-ray burst they could have estimated an a posteriori probability of this coincidence, and found it to be remarkably small. None of these was found, so these many hypotheses were rejected, and only the apparently best one, coincidence with a supernova, was put forward. The chance that *some* hypothesis (out of many possibilities) would occur by accident is much greater (a hundred million times greater in the case of the serial numbers on the dollar bill) than the chance that a *single* hypothesis, chosen in advance, would occur by accident. Perhaps the worst pitfall of a posteriori statistics is that it is impossible to say just how significant a striking coincidence is, because it is impossible to say how many other hypotheses were implicitly discarded as unpromising.

A posteriori statistical results may be valuable. Like anecdotes of unexpected observations (which is what they are), they can suggest a hypothesis for more rigorous tests in which the hypothesis is defined in advance. For example, once the possibility that some gamma-ray bursts and supernovas are associated was suggested, it was possible to compare lists of bursts and supernovas to see if they coincide more often than by chance alone. If coincidence (e.g., how close on the sky and how nearly simultaneous the events must be to qualify as coincident) is defined before the data are examined, the results, called a priori statistics, are quite reliable.

Within months of the discovery of GRB980425 and SN1998bw lists of recorded gamma-ray bursts and supernovas were compared. No other coincidences were found, beyond those expected to occur accidentally. It was possible to conclude that no more than 0.2% of bright bursts are associated with a supernova, and no more than 1.5% of the faint bursts (whose positions are measured less accurately).

Of course, this did not, and could not, disprove the reality of an association between GRB980425 and SN1998bw. It is never possible in this manner to disprove the hypothesis that a very small fraction of bursts is associated with supernovas, a fraction so small that only one association would be expected to be found in all the catalogued data. To test this hypothesis that the associations are genuine but extremely rare it is necessary to measure the positions of many (say, ten) times as many bursts as had previously been catalogued. Then about ten coincidences would be expected, though with great statistical uncertainty. Although possible in principle, this would require obtaining a large number of additional accurate gamma-ray burst positions, which won't be available for many years.

This was not the end of the story. The suggestion of an association between bursts and supernovas revived interest in the hypernova hypothesis. In fact, SN1998bw was an unusually bright and extraordinarily rapidly expanding supernova that radiated much more energy at radio wavelengths than any other known supernova. Despite the failure to confirm the statistical evidence for an association, the combination of SN1998bw's unusual properties and the possible burst association left a lingering suspicion that there was a connection, at least with a few anomalous supernovas that might be the speculated hypernovas.

There is a theoretical argument against the hypothesis that hypernovas make gamma-ray bursts (although not against the existence of hypernovas themselves). It requires the use of the concept of entropy. Entropy is often, and correctly, described as a measure of the degree of disorder of a physical system. It is also a powerful tool in information and communications theory, but its roots lie in nineteenth-century thermodynamics. Physicists have long known how to calculate the entropy of most substances, including the air around us, the hot gas in an internal combustion engine or an aircraft's turbojet, and the even hotter plasmas that make up stars. As a general rule, the entropy of a substance is higher at higher temperature and lower at higher density. Neutron star matter, extremely dense but not (by astronomical standards) very hot, has very little entropy, probably less than any other matter in the universe outside a low temperature physics laboratory.

Entropy is a powerful tool because it is usually conserved—most physical processes do not change the entropy of a gas or plasma, or change it so slowly that it can be considered constant. In this it is like angular momentum (Chapter 10), which is useful because it does not change as an object moves in its orbit, so that if its angular momentum and energy are once determined its orbit is known forever (at least until an external force changes one of these quantities, but that is very rare). Once the entropy of a gas is known, a mathematical relation connects its temperature and density. If the entropy is conserved, an increase in density implies an increase in temperature, according to a simple and well-known law, and vice versa.

Entropy is not always conserved. It changes when heat flows into or out of a substance. If you pour water on hot coals, the entropy of the water increases, and that of the coals decreases (by a lesser amount). Heat flows rapidly over short distances, but slowly over longer ones, so heat flow is unimportant when stars collapse or explode. Entropy also increases (never decreases) in shocks. However, if gas is compressed without a shock (e.g., by being pulled into a black hole), or if it expands freely (e.g., a rocket exhaust in space or the solar wind), then its entropy is nearly constant even though its density and temperature may change by very large factors.

A very massive star, such as that envisaged for a hypernova, develops at the end of its life a dense core of exhausted fuel—nuclear ashes left over from the nuclear reactions that make energy in stars—surrounded by an enormous hydrogen and helium envelope. These stars are called super giants because they are so large, and their envelopes are very dilute, with mean densities less than a thousandth that of water (or a thousandth the mean density of the Sun, which is about the same as that of water). The temperatures in these envelopes are also higher than temperatures in the Sun. This combination of high temperature and low density means that the entropy in the envelope of a super giant star is very high, much higher than in other stars.

When the central core grows massive enough it collapses to a neutron star or black hole. Woosley had suggested that a black hole would form from the core. If the star were not rotating, then the remaining matter would either be expelled in a supernova, or perhaps a hypernova (if a distinct class of such events exists), or would fall into the black hole and be swallowed without a fuss.

Rotation would change all this, and most stars are known to rotate. Rotating matter has angular momentum, which is conserved when a star collapses or explodes. Matter with too much angular momentum (not very much is required) cannot fall into the black hole because its possible orbits do not bring it close enough to the black hole to be swallowed. It can, however, settle down into a disc or doughnut rotating around the black hole.

Entropy complicates things further. Matter with very little entropy can collapse to make a neutron star. If it has too much angular momentum to do this, then it can form a rotating neutronic doughnut or a flattened neutronic pancake. However, the matter inside the massive supergiant stars assumed to make hypernovas has a great deal of entropy. This entropy keeps it from collapsing to the high density required to make a neutronic doughnut and a gamma-ray burst, for if it did collapse its pressure would be too great for gravity to hold it together. The entropy has the same effect as the air in a feather pillow: it keeps the matter fluffed up. In contrast, the entropy of actual neutron star matter, initially very low, is not expected to increase much when two neutron stars coalesce.

It is still more complicated. Once matter reaches a density intermediate between that of an ordinary star and a neutron star, say a million times the density of water, it begins to emit large numbers of neutrinos. Neutrinos can freely escape even from the centers of dense stars (they do take several seconds to diffuse out from the very densest stars, neutron stars), draining heat and entropy from the star to the cold dark emptiness of space. This is equivalent to squeezing the air out of a feather pillow to flatten it. It may be that neutrinos remove entropy from the collapsing core of a massive star (the supposed hypernova) fast enough to permit some of it to reach neutron star density and produce a gamma-ray burst.

This is one of the questions in astrophysics that can only be answered by an elaborate numerical calculation performed on a large computer. The calculation is probably feasible, but has not yet been done. However, its results may depend on unknowable details of the initial conditions—how the collapse begins. For example, such dependence has made it impossible to calculate successfully ordinary supernova explosions, despite more than 30 years of hard work. On the other hand, it may be that the hypernova collapse calculations, once they are performed, will give results essentially independent of the assumed initial conditions. Such results would be credible.

Unfortunately, even if the entropy difficulty could be overcome, making a gamma-ray burst out of a hypernova would not be easy. The burst requires the acceleration of a very small fraction (perhaps a millionth) of the star's mass to very close to the speed of light. If too much matter—and it does not take much—is mixed into the pair fireball, then the matter won't move fast enough, and won't be energetic enough, to make a gamma-ray burst. This is called "baryon poisoning"—protons and other atomic nuclei are called baryons, from the Greek word βαρνζ (heavy), because they are comparatively heavy particles—and is a source of concern even in neutron star coalescence models. The space around a neutron star is expected to be a near-perfect vacuum, swept clean of matter by its gravity and magnetic field, so there is hope that bursts produced by coalescing neutron stars might not suffer from baryon poisoning. The interior of a hypernova would be filled with matter. Baryon poisoning may be catastrophic.

Promoters of hypernova models kept hoping. They suggested that perhaps the collapsed core could make a pair fireball that would rise buoyantly through the star, pushing aside the matter rather than mixing with it, thus avoiding baryon poisoning, and finally emerging as a jet. It was impossible to decide whether nature really works this way, for the question involved, yet again, incalculable processes of turbulent mixing between the pair fireball and surrounding matter.

The observations did not help much, either. Several hypotheses could be ruled out. For example, it was clear that most observed gamma-ray bursts were not associated with *known* supernovas. It was also clear that most observed supernovas were not associated with *known* gamma-ray bursts. However, it was possible to concoct hypotheses that satisfied these observational constraints, and yet involved an essential connection between these two classes of events: For example, suppose that a hypernova creates a narrowly collimated gamma-ray burst. Only an observer who, by chance, was accurately located in the direction of the beam of gamma-rays would see the burst. In that direction the burst would be so intense it might be observed at great distances, corresponding to the observed cosmological red shifts. At those distances the hypernova itself would be so faint that it would be nearly undetectable against the gamma-ray-burst afterglow. Most observers would be outside the beam of the gamma-ray burst and would observe only the hypernova itself as an unusual variety of supernova. SN1998bw and GRB980425 would then be explained as an unusual intermediate case, in which we were close enough to the jet to see the weak outer edges of the burst (just as you will be lightly sprinkled if you stand near, but outside, the jet from a fire hose), but missed its full blast (which would have knocked the instruments off-scale at its comparatively close distance).

This hypothesis sounds artificial and contrived, but cannot yet be disproved. It predicts that a few supernovas and gamma-ray bursts will have exactly coincident positions, and that an even smaller number of supernovas will be accompanied by bursts of unprecedented intensity. When many more bursts are observed with accurate positions this prediction will be tested. The hypothesis also predicts that gamma-ray bursts are very strongly beamed, so that we miss most of them entirely, even though we see the accompanying hypernovas. Then the total rate of bursts must be much greater than if we observe them all, or a larger fraction of them. This means that the apparent rough consistency between the observed burst rate and the neutron star coalescence rate must be a fluke, rather than confirming the coalescence model, but this was already assumed when it was hypothesized that hypernovas, rather than coalescences, make bursts.

Supernovas like SN1998bw are known (because the mass of exploding matter can be inferred from the observed brightness and spectrum) to be the product of the deaths of very massive stars. This accorded with the hypernova hypothesis. Such stars are short-lived by astronomical standards (a few million years), which led to the suggestion that gamma-ray bursts would be found in regions of galaxies in which massive stars are still being formed, or formed in the comparatively recent past. If gamma-ray bursts were the product of coalescing neutron stars this might not be the case, for neutron stars are believed (on the basis of somewhat uncertain evidence) to form from less massive (about ten solar masses) stars, which live longer. In addition, the known double neutron stars will not coalesce for hundreds of millions of years, although there is an obvious bias in observing such systems—those whose orbits coalesce more rapidly have that much less opportunity to be observed.

In principle, therefore, observations of the locations of gamma-ray bursts could determine whether they are produced by the coalescence of a binary neutron star or the collapse of a single very massive star in a region of ongoing star formation. In practice, a galaxy at the known distances of gamma-ray bursts is a tiny and very faint fuzzy patch of light, even when observed with the Hubble Space Telescope, and it is difficult to tell what kinds of stars make up the region from which the burst, or its afterglow, is observed. It might be possible to make indirect arguments based on the fact that active star-forming regions have a great deal of dust that absorbs visible light and would dim the afterglow, and comparatively dense interstellar gas, but no clear consensus emerged.

The shaky case for hypernovas was somewhat strengthened when X-ray astronomers found evidence for large quantities of iron in the spectra of at least three bursts. The evidence was too strong to dismiss easily, but not quite strong enough to be compelling. It suggested that a supernova or stellar wind ejected a large quantity of matter, perhaps enriched in iron (which some supernovas, and probably also hypernovas, make in abundance) into the region surrounding the burst. However, this ejection must have occurred a significant time, probably between 1 and 10 years, before the burst itself. This did not fit neatly into any hypothesis. Coalescing neutron stars might be expected to have been produced by supernovas millions of years ago. The hypernova hypothesis predicted an explosion simultaneous with the burst, not years earlier. As too often in astrophysics, there was ample room for speculation, but few ways to test the speculations.

16

The Holy Grail

A few miles from Los Alamos, New Mexico, stands one of the most remarkable astronomical observatories in existence, the Robotic Optical Transient Search Experiment (ROTSE). It is completely automated, steered by electronic signals received over the Internet. Housed in a military surplus hut (purchased from a scrap dealer) the size of a large closet, it consists of a cluster of four commercial telephoto camera lenses (200-mm focal length, $f/1.8$), each about 4 inches in diameter. Made by Canon, they can fit on an ordinary Canon 35-mm camera. At $4199 apiece (in 1995), they were too expensive for most amateurs, but just right for the professional sports, nature, or news photographer. Each lens feeds light to a CCD (charge-coupled device, an integrated circuit that converts a visible image to electrical signals) camera, very similar to the CCD in an ordinary digital camera. Each CCD divides its image into about four million picture elements, an impressive figure in 1995 but one that consumer digital cameras are now approaching. An obsolescent 133 MHz PC, running the Linux operating system (for reliability Windows just won't do) controls each CCD. If it weren't bolted to a mount, you could pick up the entire telescope and camera array and carry it off under your arm. On January 23, 1999, one of these four cameras recorded visible light from a gamma-ray burst as it was happening, which had been the holy grail of gamma-ray-burst astronomy for a quarter of a century.

From the discovery of gamma-ray bursts, astronomers had asked themselves if the gamma rays were accompanied by visible light, and if this light could be detected. Because there was no theoretical understanding of bursts, or even a model that could be calculated in detail, it was not possible to predict how bright their visible counter-parts would be. However, if even a tiny fraction of the gamma-ray energy appeared in visible light, they would be quite bright. Analogy to known bursting X-ray sources suggested that this fraction might be between 0.1% and 1%. If gamma-ray bursts were found in binary stars, one of the most popular early theoretical ideas, gamma rays would be absorbed in the atmosphere of the companion star and roughly 0.1% to 1%, the exact percentage depending on the distance to the companion, its size and properties, and the gamma-ray inten-sity, reradiated as visible light. In fact, this is exactly what happens in X-ray bursters.

This is not a very large fraction—you hardly notice it when the sales tax is raised by 0.1%, and most investors happily accept that the managers of their mutual funds and other investments rake off about 1% of their assets every year in fees and expenses. Yet if 1% of the energy of a bright, but not extraordinary, burst (a "burst of the month") were converted to visible light, it would be about 6th mag-nitude. That is visible by a good naked eye in a dark sky, and is ex-tremely bright by the standards of professional (or even serious am-ateur) telescopic astronomy.

Of course, you would have to know when and where to look. Ay—there's the rub, because gamma-ray bursts are unpredictable. There are two possible approaches to this problem. One is to use the direc-tion to the burst, determined by the gamma-ray observations, to steer the optical telescope. Unfortunately, data analysis was slow, and po-sitions measured by the Vela satellites, and the later interplanetary networks, did not become available until weeks or months after the bursts. The other approach is to design an optical system, using a form of fish-eye lens, that collects light from as much of the sky as possible. Then, if a flash were detected and recorded, the optical data could be compared at leisure to gamma-ray data to see if they occurred in the same place on the sky and at the same time. If no flash were seen, it would at least be possible to set an upper bound on how much visible light was emitted by any gamma-ray burst in the portion of sky under observation.

There was also the hope of discovering some completely new phe-nomenon that might make flashes of visible light alone, without gamma rays. For example, an event like a gamma-ray burst, but with lower Lorentz factor (more baryon poisoning), might radiate most of its energy as visible light rather than gamma-rays. Or, there might be something completely unrelated to gamma-ray bursts. New phenom-ena are generally discovered by new instruments with new capabilities. That is how gamma-ray bursts themselves were discovered, as well as pulsars, radio galaxies, the cosmic background radiation, and even the expansion of the universe. Fame and prizes, the chief motivators of scientists, are the rewards for discovering something really new.

The performance of any optical system is governed by certain laws, the most important of which is called Liouville's theorem. In essence, it states that if an optical system (a telescope or a camera) is to have a large field of view, the range of angles from which it collects light, it must be small. A telephoto lens, with a comparatively small field of view, may be larger than a standard camera lens, which in turn is larger than a wide field of view (fish-eye) lens. An ordinary astronomical telescope is really only a camera with a very large telephoto lens, or a mirror that takes the place of a lens.

Just how small an optical system must be depends on the size of the device that records the light, onto which the light is focused. Modern astronomical systems use CCDs because they are very efficient, recording nearly all the visible photons falling onto their surface (in contrast to photographic emulsions, which record less than 1%), and because their electronic data are easy to process by computer. Unfortunately, CCDs are generally no more than an inch square.

For the astronomer looking for faint objects, this means that if he is to detect them over a large swath of sky his collecting lens or mirror must be small. It won't collect very much light, and the instrument won't be very sensitive. In addition, each picture element of his detector will receive light from a broad area of sky, including whatever stars and stray light there are, making it difficult to observe faint flashes against this background. The more picture elements into which he can divide his field of view, the less of an obstacle the background will be, and the more sensitive the instrument will be. Increasing the number of picture elements is the chief goal of CCD manufacturers; if it were easy, it would have been done already. This is the reason ROTSE uses four separate lenses and CCDs, for a total of sixteen million picture elements. More would have been even better, but the budget was limited.

If the entire sky must be monitored for possible visible bursts, the astronomer needs as big a field of view as possible. It is not necessary that he detect every burst, and he cannot come close. Some will be below the horizon, or in the haze near it. Others occur in daytime, or twilight, or in moonlight (a serious source of background light), or in bad weather, or behind the thick clouds of dust and soot that fill the plane of our galaxy. Combining these factors means that, at best, a single observatory can see about 3% of the bursts that occur, or about one per month, most of which are weak. The astronomer cannot afford to lose many more, so his instrument must have a field of view that encompasses most of the sky. Liouville's theorem then implies that his collecting lens cannot be larger than the CCD that records the data, perhaps an inch in diameter, and maybe even smaller. He must do research with a lens smaller than that in a child's toy telescope. Multiple lenses and CCDs help, but also multiply the cost.

This sounds hard, and it is. The first proposal to monitor the sky for optical counterparts of gamma-ray bursts was made by Paul Boynton of the University of Washington in 1974, not long after the bursts themselves had been discovered. Boynton was trained as a physicist, but moved into astrophysics, studying pulsars, especially those that flash in visible light, and was uniquely well suited to search for optical transients. CCDs were not yet available, so he proposed to use the best technology of the time, a vacuum-tube imaging device called a vidicon. It would have only 25,600 picture elements, but he still predicted it would be able to detect 10th magnitude flashes lasting a second. In fact, he planned to use two telescopes some miles apart as a stereo camera, so that he could tell whether a flash was nearby (a meteor, or the sun reflected by an artificial Earth satellite) or distant, like a gamma-ray burst. He also planned on completely robotic operation, necessary to keep the costs reasonable, and every subsequent proposal has followed his lead. Boynton's proposed instrument would have been more than sufficient to detect the 6th magnitude flashes suggested by a naïve guess of 1% gamma ray to optical conversion. In fact, it might have detected a 9th magnitude flash like that finally observed in 1999. Promises of instrumental sensitivity have a history of being overly optimistic, but the development of CCD technology would have improved the sensitivity over the original design.

After working out the parameters and rough design of his detector system, and publishing them in the proceedings of a conference, Boynton submitted a formal proposal to the National Science Foundation for support to pay for its construction. The NSF rejected his proposal, citing a referee who said he had "failed to show that [they] would, in fact, observe anything." Of course, if they had known in advance what they would discover, it would not be a discovery. The proposed instruments were unprecedented, and only possible because of advances in technology. It was easy for unimaginative reviewers to attack such a proposal. The reviewing process invites attacks, and even one negative opinion out of five or six reviews is usually sufficient to ensure rejection. Had this proposal been approved, the visible counterparts to gamma-ray bursts might well have been discovered 20 years earlier than they actually were, and the nature of the bursts understood much sooner.

Boynton gave up on gamma-ray bursts, and went on to a successful career in X-ray astronomy and fundamental physics. The subject of optical counterparts went to sleep until the early 1980s, when Schaefer, examining archival photographs, reported finding bright transients at the positions of bursts, but decades before the bursts themselves (Chapter 8). This revived interest because his results seemed to imply optical counterparts of 6th magnitude or brighter, which appeared readily detectable. A number of instruments were proposed, and some of them actually built. The best known were the Explosive Transient Camera (ETC), developed by a team led by George Ricker at MIT, and the Rapidly Moving Telescope (RMT), a project of the NASA Goddard Space Flight Center under the leadership of Bonnard Teegarden. Preliminary designs and plans were announced in 1983.

These two instruments were designed to work together, automated and unattended. The ETC would stare at a large swath of sky, waiting for flashes of light. Once a flash had been detected, its position would be transmitted to the RMT, a much larger telescope (originally planned to be 7 inches in diameter, compared to about 1 inch for each of the sixteen individual lenses in the ETC), with a small field of view and greater sensitivity, which would steer to the position of the flash in a few seconds. Astronomical telescopes usually have plenty of time to move from one target to another, which is called slewing, because most astronomical objects are permanent, available to be studied whenever convenient for the astronomer. A telescope will generally point at one target for many minutes at a time, gradually accumulating light from a faint star or galaxy, and only moving to follow the rotation of Earth (which makes everything in the sky rise and set like the Sun). Slewing to the next target in a hurry is not usually important.

Rapid slewing would be essential to the RMT. By turning to the position of a flash in a few seconds or less it would catch a burst as it happened, collecting much more accurate data than the tiny ETC. The ETC, detecting the visible counterpart of the burst and providing approximate but timely coordinates, would take the place of a gamma-ray burst detector in space that could determine burst coordinates and radio them to an optical telescope, in "real time," while the burst was still going on. The RMT would also produce a sharp image of the transient. This would not show any detail—gamma-ray bursts were much too distant, in anybody's model, for that. It would give a precise position, accurate to about 2 arc-seconds, which could later be used to steer a large telescope to the position of the burst and see what was there, perhaps a faint star in our galaxy or a distant galaxy.

The ETC, originally planned to be operational by 1985, was a long time coming. Funding was limited (the NSF also rejected a proposal to support its development). Many novel technologies, particularly information-processing algorithms and communications protocols, needed to be worked out. One ETC finally began collecting data in 1991, a few months before the launch of GRO and BATSE. When finally completed, ETC had a much smaller field of view than originally planned. Instead of staring at 43% of the sky above the horizon, it only stared at 12% (at an intermediate stage in its development the figure was down to 6%). This nearly fourfold reduction meant a corresponding reduction in the rate at which gamma-ray bursts would occur within its field of view. Making the usual allowances for daytime, weather, galactic absorption, twilight, and moonlight meant that it could only observe about ½% of whatever flashes there are. Either because this fraction was so small, or because it was not sensitive enough (it was estimated to be able to detect 1 second flashes as faint as 7th or 8th magnitude, although it was originally hoped to be able to detect flashes down to 11th magnitude, more than twenty times fainter), ETC never found any convincing evidence for visible flashes.

ETC found an enormous number, hundreds per night, of spurious flashes, mostly sunlight reflected by artificial Earth satellites, meteors, clouds scattering moonlight, or stars appearing from behind clouds. Had there been two ETCs, operating as a stereo pair as originally planned, the spurious flashes could have been eliminated in real time, and the positions of any genuine flashes (as well as the few spurious flashes not so easily eliminated) handed off to the RMT. Because there was only one ETC the discrimination of spurious flashes could only be done "off-line," after some delay, too late to save the RMT from chasing a large number of spurious events. It may be possible to find a needle in a handful of straw, but ETC was giving the RMT a whole haystack. As a result, the RMT never collected useful data.

A fortuitous failure on the Gamma-Ray Observatory soon made ETC, and all similar sky-staring telescopes, obsolete. As originally designed, all instruments on GRO, including BATSE, would record their data on an onboard tape recorder. Several hours of data would be accumulated and then transmitted to the ground over NASA's system of Tracking and Data Relay Satellites (TDRS). Not long after the launch of GRO its two tape recorders (the principal one and its backup) began to fail. This became worse and worse, until they became unusable in early 1992, before the satellite had been in space for a year.

It would have been easy simply to declare GRO a complete loss and shut it down, but this time NASA did the right thing. A new ground station for the TDRS system was built and installed, first in Guam and later in Australia. This enabled data to be relayed to the ground as they were received (in real time), without any significant delay. The scientists could see the data coming in as a gamma-ray burst was happening.

Scott Barthelmy of NASA's Goddard Space Flight Center recognized this as an extraordinary opportunity. The BATSE data included not only the brightness and spectrum of a burst, but also its position on the sky. By design, and because of the technology used in BATSE, these positions were very approximate (the error circles were believed to be between 4 and 10 degrees in radius), but they were accurate enough to permit a ground-based telescope to be pointed in that direction. In a stroke both ETC and RMT became obsolete, ETC because BATSE was now providing gamma-ray-burst positions directly, without depending on the initial detection of an optical flash and verification of it as a genuine cosmic event, and RMT because its field of view was much too small to view more than a tiny fraction of the entire BATSE error circle.

Barthelmy seized this opportunity by constructing the BATSE Coordinates Distribution Network (BACODINE, later renamed the Gamma-ray burst Coordinate Network, or GCN) to distribute the information from BATSE. He did this entirely on his own, without funding from NASA, scrounging and "bootlegging" resources as necessary. When completed, NASA management was amazed how quickly and economically it was done, for had it gone through a formal planning process it would have cost several hundred thousand dollars and taken much longer.

By the middle of 1993, about 8 months after the original idea, BACODINE was up and running, calculating burst positions from BATSE data and distributing them over telephone lines. In a few more months Internet and e-mail distribution was added. Any astronomer or observatory, anywhere on Earth, could now learn the coordinates of a burst within about 5 seconds of its detection by BATSE. The majority of bursts would still be going on.

Barthelmy's ambitions were not limited to distributing coordinates. He realized that the coordinates, by narrowing the field of view that needed to be monitored, would make the detection of a simultaneous visible counterpart much easier. Instead of looking at as much of the sky as possible, it would be sufficient to slew a telescope to the position indicated by BATSE. The required field of view would still be very large by astronomers' standards (8 to 20 degrees across), but much smaller than that required to stare at the entire sky. The instrument could have a much larger lens that would collect more light, and it could be much more sensitive. Essentially, it would be a hybrid of the ETC (itself following Boynton's 1974 design principles) and the rapidly slewing RMT, with a field of view and optical design intermediate between these two instruments. It could be thought of as a modified RMT using BATSE and BACODINE in place of the ETC.

He called his proposed instrument the Gamma-ray to Optical Transient Experiment (GTOTE). Together with BACODINE, it might have enabled him to discover the first simultaneous optical counterpart of a gamma-ray burst. Unfortunately, it was not supported by NASA and was never completed.

Gamma-ray-burst astronomers were not the only scientists needing wide field-of-view optics. In the 1980s the U.S. Strategic Defense Initiative, popularly known as Star Wars, was looking for ways to detect and track missile launches and reentering warheads from space. Brilliant Pebbles was a scheme to destroy enemy rockets and warheads by smashing a solid body into them (an earlier version had been called Smart Rocks) and it needed accurate tracking. One of the methods considered was optical imaging. A wide field of view would be required because a threat could come from a broad range of directions.

The Lawrence Livermore National Laboratory is a large institution (with about 7000 employees, roughly half of them scientists, engineers, and programmers) run by the University of California in Livermore, California, 40 miles east of San Francisco. Its chief mission is nuclear weapons (it is a sister to the Los Alamos National Laboratory, where the first atomic bomb was developed), but it also engages in many other kinds of defense research, in addition to a substantial program of basic research unrelated to defense. In the late 1980s Livermore received a contract to develop a wide field-of-view camera for space defense. Hye-Sook Park, trained as an experimental particle physicist, led its development. Brilliant Pebbles then ran into trouble, and Livermore's wide field-of-view camera gathered dust.

Carl Akerlof is an experimental particle physicist at the University of Michigan (where Park had been a student). During the 1970s and

1980s particle-physics experiments grew to require ever larger teams, in some cases consisting of several hundred scientists. This reduced the independence and opportunities for initiative of all participants, and he looked to observational astrophysics for science on a smaller and more human scale. He became involved in an experiment (Whipple) observing high-energy gamma rays, using Earth's atmosphere as a detector, a technique very similar to those of particle physics, and in MACHO, which uses optical telescopes to study dark matter in our galaxy by observing gravitational focusing (microlensing) of the light of distant stars by the dark matter.

In 1992–93 Akerlof went to Berkeley for a year's sabbatical (a temporary appointment on the faculty of another institution) because the MACHO project was led by Livermore with a large Berkeley contingent. He went to Livermore to visit Hye-Sook Park, whom he had known slightly from her student days. Akerlof had earlier become interested in the problem of searching for optical counterparts of gamma-ray bursts, and had heard of the Livermore wide field-of-view camera (even though it had a defense application, it was not classified and some details had been published). He was pleased to discover that Park not only knew about it, but was able to show it to him, "abandoned and unloved . . . as they opened the enclosure to see the control electronics, spiders scurried out of sight behind the printed circuit boards." It was clear that this was the right instrument to begin a search for the optical counterparts of gamma-ray bursts, and a collaboration was born. Livermore management was happy to provide funding. Park, her programmers and engineers, and Brian Lee, a University of Michigan graduate student, brought the wide field-of-view camera back to life as the Gamma-Ray Optical Counterpart Search Experiment (GROCSE).

Unfortunately, GROCSE was not very sensitive. This had not been a problem for its original mission as part of Brilliant Pebbles, for rockets and reentering warheads are rather bright, but was a serious difficulty in gamma-ray-burst astronomy. It could detect a 1-second flash as faint as 8th or 9th magnitude, but no dimmer. This was perhaps a little better than ETC, but not a great improvement.

GROCSE had the advantage that by responding to BACODINE alerts it was at least sure to be pointing in the right direction. ETC, in the words of its builders, did "not require a trigger from BATSE or any other experiment," a backhanded way of saying it did not take advantage of the information distributed by BACODINE that would have told it where to look to find a burst. They never adopted the strategy of rapid slewing planned for GTOTE and used by GROCSE (and its successors). ETC only gave useful information if the burst happened to be within its predetermined field of view, which covered about an eighth of the sky. It would be looking the wrong way during seven-eighths of the bursts.

Like ETC, GROCSE did not detect any bursts. Their upper limits were sufficient to disprove the naïve assumption that 1% of the burst energy was converted to visible light. However, this assumption was no longer relevant; the binary models that had led to it had been disproved by the BATSE statistical data demonstrating that the bursts must be at cosmological distances. The lower assumption of 0.1% conversion was still permitted by the data.

It was clear that more sensitivity was needed. Sensitivity could be improved in two ways. First, the old Star Wars camera had been designed to produce rapid series of images, not to study faint objects. A new instrument, optimized to do astronomy, would perform much better. CCD technology had improved dramatically since the wide field-of-view camera had been built. The GROCSE CCDs had only 221,184 picture elements each, while state-of-the-art CCDs had 4,194,304, nearly twenty times as many. GROCSE's complicated optical design was also rather "slow," in the terminology of camera and telescope designers, meaning that the lens diameter was comparatively small and therefore did not collect much light.* GROCSE also used a complicated system in which light was first passed through fiber optics and then amplified by an inefficient and noisy vacuum-tube device called an image intensifier before it reached the CCD.

Second, the BATSE gamma-ray-burst coordinates, transmitted through BACODINE, made it possible to reduce the field of view. It was only necessary to look at the patch of sky that might contain the burst, rather than at as much of the sky as possible. By Liouville's theorem, this permitted larger lenses that would collect more light. It, along with the vastly greater number of picture elements on improved CCDs, also meant that each picture element would be smaller, so there would be less starlight and skylight in it to overwhelm the faint hoped-for signal of a gamma-ray burst. An optical system with a field of view (about 16 degrees across) matched to the uncertainties in the BATSE positions could have a sensitivity hundreds of times

---

*The technical term for this is that its f-number, the ratio of focal length to lens diameter, was 2.8, as compared to values between 1.4 and 1.8 for the 35-mm camera lenses used by amateur photographers. ETC originally proposed to use lenses with an f-number of 0.85, but wound up with 1.4. Lenses with smaller f-numbers collect more light and are called "faster" because they permit shorter exposures.

greater (about 14th magnitude). In essence, BATSE would replace ETC. RMT had much too small a field of view to fit the BATSE coordinates, so it would be replaced by a new instrument, initially called GROCSE-II.

New instruments cost money. GROCSE-II would be small, but it would be custom-designed and built, and would advance the state of the art in robotic telescope control and data processing. There were internal funds for work at Livermore, but Akerlof needed support for his work at Michigan. In 1994–95 he submitted a total of four proposals to the NSF. In each case he received excellent reviews, along with a form letter of rejection. GROCSE-II nearly died. Apparently, each dollar of the NSF astronomy budget had someone's name on it, and there was no room for new people or ideas, even those acknowledged to be original and excellent. He received a little support from NASA, some from an internal University of Michigan research fund, and a crucial grant from the Research Corporation, a private philanthropy not bound by the bureaucratic constraints that hobble the NSF.*

Scientists are notorious for squabbling about credit for discoveries,† and sometimes they simply don't get along. The development of GROCSE-II was well under way in early 1996 when an ugly split developed between Akerlof and Park, its two leaders. What may have begun as legitimate differences of opinion soon became a struggle for control. Divorces, in science as well as in marriage, generally involve irreconcilable differences, and usually the parties involved give irreconcilable accounts of what went wrong.

According to Akerlof, Park decided to cut him out of the project, even though it had been his idea to turn the defunct wide field-of-

---

*The real reasons for NSF funding decisions are hidden behind their form letters. If a disappointed applicant inquires, he will be told of the large number of excellent proposals, but given no insight into how the hard decisions are made. If he suspects favoritism, cronyism, or just closed minds, none can prove him wrong. It appears that the NSF is much more interested in big science than in small science, even though new ideas start small; it may be as hard to get two million dollars as two hundred million, and not much easier to get fifty thousand. The big hogs push the piglets away from the trough, and sometimes they eat the piglets.

†One prominent theorist had the curious habit of sidling up to a younger scientist who had just presented his ideas and saying that he, too, was working on that subject, and that they should write a paper together. The speaker, flattered or perhaps intimidated, would agree, but when the paper was finally written the prominent scientist would have contributed only his name.

view camera into the functioning GROCSE instrument. Akerlof and Lee had played a major role in making GROCSE work, as well as in designing GROCSE-II. Park denied him and the Michigan team access to the GROCSE hardware, software, and data. He describes her acts as amounting to "theft of intellectual property," and considered legal action.

Neither Park nor anyone else at Livermore was willing to give her side of the story. Akerlof can be blunt and outspoken. When I telephoned him, his first statement was "I suppose you want me to write your chapter for you." (I did not and he did not.) This may not make him easy to work with (he withdrew from both the Whipple and the MACHO experiments after some friction), but bland organization men do not make scientific entrepreneurs. On technical matters he is usually right.

Eventually, a settlement was reached. Livermore made a cash payment to the University of Michigan and agreed to support a staff member working on ROTSE. The existing equipment was divided between the two teams, something that Akerlof compared to Solomon dividing the disputed baby. Fortunately, this separation was not fatal, but resulted in two similar but competing experiments. Livermore may have gotten the better end of the settlement, because Park's Livermore Optical Transient Imaging System (LOTIS) was in operation before the end of 1996. Akerlof's ROTSE moved to Los Alamos, again illustrating that the U.S. nuclear weapons laboratories are readier to provide venture capital for new scientific ideas, even those with no connection to weapons, than the government agencies charged with their support. ROTSE was in operation by early 1998.

The GROCSE data, reporting upper limits to the brightnesses of possible optical counterparts to gamma-ray bursts, were published twice, first by Michigan and then by Livermore. Much of the work, especially the data analysis, had been done by Brian Lee. Akerlof was careful to ensure that Lee's name appeared first on the author list (usually a mark that this author bears chief responsibility, and should receive most of the credit, for the work) of the Michigan paper (none of the Livermore people were listed, at their request), but Park's name was first on the Livermore paper (no Michigan people other than Lee were listed).* Quite properly, the BACODINE team were on both au-

---

*ROTSE papers had the authors listed in alphabetical order, a solution to the problem of squabbling over their order, but one that left the reader wondering if Akerlof was first only because his name begins with the letter A.

thor lists. Multiple publication of the same results is usually strongly disapproved of because research papers are supposed to report only *new* results, but was tolerated in this case.

The theorists had not been entirely idle, and had tried to make more sensible predictions of the brightness of burst counterparts. The development of relativistic shock models of bursts had led in 1994 to a prediction that at frequencies below the gamma-ray range the brightness (Chapter 14) would vary as the ⅓ power of frequency. Extrapolated down to visible frequencies, it meant that less than a millionth of the energy of a gamma-ray burst would appear in visible light. A bright "burst of the month" would be approximately 18th magnitude. This was very discouraging, because it was far below the sensitivity of ROTSE or LOTIS. Much larger telescopes, with smaller fields of view, would be necessary. The BATSE coordinates would not be accurate enough to point these telescopes, and they would have to wait for future space instruments that could locate gamma-ray bursts more accurately.

Fortunately, the theorists did not stop there. Re'em Sari and Tsvi Piran looked more closely at the physics of the shock produced when a relativistic debris shell collides with the interstellar medium or other dilute gas. There will actually be two shocks, one in each fluid, just as when you clap your hands together both hands sting. If one fluid is much denser (the debris shells may be perhaps a million times denser than the interstellar medium when they collide), the shock in it is much weaker, just as if your hand slapped a boulder rather than your other hand; your hand may be hurt, but the boulder will not be. The shock in the dilute fluid is strong, and makes electrons energetic enough to radiate gamma rays. The shock in the dense fluid is much weaker. It won't radiate gamma rays, but it can radiate visible light. The spectrum of the gamma rays cannot be extrapolated to visible light, because the visible light is produced by a different source with different properties. Sari and Piran predicted that the visible counterparts would be much brighter than simple extrapolation from the gamma-ray burst had implied. Their results depended on many uncertain parameters, so that it was impossible for them to be very specific, and the visible to gamma-ray ratio would probably be very different in different bursts. These predictions were presented at a meeting held in Rome in November 1998, and the paper describing them was distributed electronically to the worldwide astronomical community on January 10, 1999.

By this time LOTIS had been in operation more than 2 years, and ROTSE nearly a year. Each had received via the GCN (formerly

BACODINE) scores of gamma-ray-burst positions within seconds of the beginnings of the bursts. If conditions were favorable (night and good weather at the observing site, and the burst above the horizon), the position of the burst was observed. The analysis was slow because the cameras recorded data over the entire large BATSE error circles, which had to be searched for a possible optical transient. It was easier if BeppoSAX or the interplanetary network obtained an accurate burst position. Even though the accurate position only was calculated after some delay, it could be used to guide the search of the LOTIS and ROTSE images obtained during the bursts. The data had to be obtained during the bursts themselves, but there was no hurry doing the analyses.

At first, only upper limits were found. In the best cases these limits, obtained during the bursts, were 13th magnitude or brighter. Data accumulated over several minutes to an hour led to even tighter bounds, as faint as 16th magnitude, but properly these were only bounds on the early afterglow, because the bursts themselves were long since over.

On January 23, 1999, came the breakthrough. BATSE detected a very strong burst, called GRB990123. Within seconds, GCN transmitted its position to astronomers and telescopes all around the world. It was raining at LOTIS, in the hills east of Livermore, but clear at ROTSE in New Mexico; perhaps it was fortunate that Park and Akerlof had split, because there were now two instruments, far enough apart that they had different weather. ROTSE steered to GRB990123 and, for the first time ever, detected a gamma-ray burst with visible light as it was happening. It was unexpectedly bright, 9th magnitude at its peak. The data are shown in Figure 16-1.

Ninth magnitude astounded almost everyone. It was much brighter than the upper limits on other bursts seemed to imply, even allowing for the greater gamma-ray intensity of GRB990123. Naïve extrapolation from the gamma-ray spectrum might have predicted about 16th magnitude (brighter than the 18th magnitude originally estimated because this burst was so unusually intense, and because of uncertainties in the extrapolation procedure itself), but it turned out to be a thousand times brighter still. The prediction made by Sari and Piran a few months before, and published electronically only 2 weeks earlier, was confirmed. In fact, the conclusion that the ratio of visible light to gamma rays is higher in some bursts than in others also agrees with their suggestion that the visible brightness depends on several parameters in a complex manner.

Observing a single event of a class is revealing, but it is also tan-

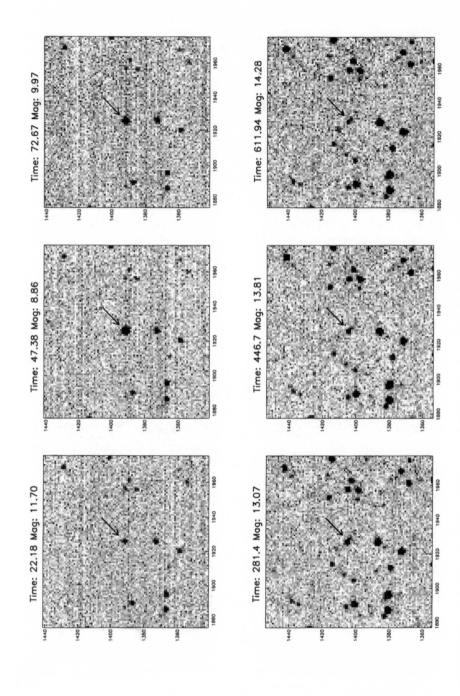

talizing and frustrating. We do not know how typical GRB990123 was, or how bright the typical gamma-ray burst is—just below the threshold of detection by LOTIS and ROTSE, or much fainter? These questions will only be answered when the visible counterparts of more bursts are observed. More surprises may be waiting. The holy grail of gamma-ray-burst astronomy has been found, and drinking from it will become routine.

Figure 16-1. Visible light images of GRB990123 obtained by ROTSE. The arrow points to its visible counterpart. The times are measured in seconds and counted from the beginning of the gamma-ray burst as recorded by BATSE. The magnitudes are also indicated, with the smallest number (8.86) in the second frame when the visible counterpart of the burst was brightest. This also corresponded to the peak gamma-ray intensity. The upper row of images was 5-second exposures and the lower row was 75-second exposures, explaining why the background stars appear so much darker (brighter in these negative images) in the lower row. The axis labels are picture elements in the CCD, whose edge occurs at 2048; ROTSE nearly missed this burst, which occurred near picture element 1925 on the horizontal axis! (Reprinted by permission from *Nature* V. 398 p. 401 © 1999 Macmillan Magazines Ltd.)

17

The End of the Beginning

The end of the Gamma-Ray Observatory and BATSE in June 2000 put a temporary stop to the search for optical counterparts, because no longer would ROTSE and LOTIS be alerted when a burst was happening. They would still study variable stars, comets, and asteroids, for which their wide fields of view make them powerful tools, as they had whenever not alerted by GCN that BATSE was observing a burst.

This will not be the end of gamma-ray-burst astronomy, nor of the study of their visible counterparts. New satellites have been designed with that search in mind, and GCN remains ready to transmit burst coordinates to observers everywhere. The first of these, HETE-2 (High Energy Transient Explorer-2, succeeding HETE, which did not make it into orbit because of a rocket failure in 1996) was launched in late 2000, and Swift (not an acronym!) is scheduled to follow in 2003. They will determine much more accurate gamma-ray-burst coordinates than BATSE. ROTSE and LOTIS will become obsolete for gamma-ray-burst work. They will be succeeded by ROTSE-II, ROTSE-III, and Super-LOTIS. These are much larger and more sensitive instruments (though with diameters of 18 and 24 inches they are still almost amateur-sized telescopes), designed with fields of view to match the accuracy of the gamma-ray-burst positions provided by the new satellites.

Scientific problems have a natural life cycle. They are born with an unexpected discovery, such as the observation of gamma-ray bursts by the Vela satellites. This immediately calls for explanation. The information available is preliminary and incomplete, so new measurements must be made. Theoretical models are floated and tested against our understanding of the laws of nature and against the data.

Sometimes a satisfactory explanation is quickly developed. Sometimes closer examination shows that there is really nothing to explain, for the original data were spurious or misinterpreted.

It is more interesting if the discovery is genuine, and remains enigmatic, as was the case for gamma-ray bursts. Then the next stage consists of the development of new instruments to answer the experimental questions, and of new theoretical ideas to explain their results. Beginning in the mid-1970s gamma-ray-burst instrument development proceeded along two lines. One was interplanetary networks, systems of detectors distributed across the solar system in order to determine accurately the positions of a few bursts. The other was BATSE, designed to find approximate positions for a large number of bursts.

Interplanetary networks functioned as planned, but unfortunately their results did not solve the gamma-ray-burst problem. No visible counterparts were found at the positions of most well-localized bursts. Others apparently led to archival optical counterparts, which have never been satisfactorily explained, but which are most likely artifacts. One burst came from a young supernova remnant in the nearby Large Magellanic Cloud—a genuine discovery, but one that turned out to have nothing at all to do with bursts in general.

The theorists were thwarted by the uncertainty as to the distances to gamma-ray bursts. Most of them, if they worked on the problem at all, preferred to assume that the bursts were comparatively close to us, within our galactic neighborhood, partly because that appeared to make the task of explaining them easier and partly because of data (spectral, archival counterparts, and the supernova remnant in the Large Magellanic Cloud) that later turned out to be spurious or irrelevant.

Progress only came after the launch of BATSE in 1991. Its data quickly persuaded almost all astronomers that gamma-ray bursts were extremely distant and extremely luminous. Suddenly, all the pieces began to fit together. The theorists even managed to develop a rough outline of how bursts work (at least, they persuaded themselves they understood them), and succeeded in predicting a number of their properties, including their afterglows. Discovery of the afterglows led to direct proof and measurement of the cosmological distances of the bursts.

This completed the second stage in the life of a scientific problem, in which new experiments or observations, specifically designed to address it, lead to a generally accepted rough working model. However (to paraphrase Winston Churchill after the victory at El Alamein), this is not the end, or even the beginning of the end. Rather, it is only the end of the beginning.

In some scientific problems, such as pulsars and quasars, this stage is over in a matter of months. In others it takes much longer. For example, the solar neutrino problem is only beginning to be understood more than 30 years after the first experiment failed to observe the predicted number of neutrinos, and the missing mass in the universe has not been found 65 years after its gravitational influence was discovered. For gamma-ray bursts, the second stage lasted 25 years from their discovery.

The third stage in the life of a scientific problem involves the development of a more quantitative and predictive understanding. This is more often successful in other branches of science than in astronomy, because in most experiments it is possible to control the initial conditions, while in astronomy they are beyond our control, and usually beyond even our knowledge. At present, gamma-ray-burst astronomers are busy trying to determine the degree of beaming of the fireball debris, its energy, and the density of the surrounding medium from the properties of their afterglows. It is unclear whether this is, in fact, possible, or simply an exercise in fitting parameters that may have little real physical significance. The scientists have some of the tools—a basic model of bursts and their afterglows—but not others—a detailed fundamental understanding of collisionless shocks and the acceleration of energetic particles—and it is unclear whether they will be able to finish the job.

The life of a scientific problem, like all lives, ends finally in death. The problem is understood as well as anyone is interested in understanding it, or it may be beyond human understanding. Known experimental methods and theoretical ideas have yielded all they can. It ends in triumph, or in frustration, or in despair, or simply in boredom. The last is probably the most common. The excitement of discovery is not replaced by routine science, as some philosophers and historians of science would have us believe, but by the silence of the grave, as scientists move on to something else, new and more interesting.

Afterword

The study of gamma-ray bursts exemplifies the progress of science in general. Fundamental discoveries are generally serendipitous. They occur, completely unexpectedly, in the course of some quite different endeavor, such as the discovery of gamma-ray bursts by instruments monitoring the Nuclear Test Ban Treaty. Even when a discovery is the result of a deliberate search, such as Watson and Crick's determination of the double-helix structure of DNA, usually the most important part of the discovery is something unexpected. For example, the importance of the double helix was how well it explained DNA replication, genetics, and heredity, which went far beyond the planned objective of simply determining the structure of a molecule. Only when theory is in advance of experiment is a great discovery the fulfillment of a specific prediction. Theory is rarely so powerful or successful.

In Chapter 2 we saw that experimental science is largely the product of technology. Theory is only the concentrated essence of experiment; it is the compression of a great body of experimental experience into a few compact rules. Without technology there could be no experiment, and without experiment there could be no theory. There could only be mathematics and philosophy, and they are not science.

Scientists watch the ebb and flow of ideas, the large and small discoveries that gradually, or suddenly, change how we think about the world. Occasionally, we turn to the writings of a philosopher of science to see if he has anything interesting to say. Usually, we conclude that he is playing games with words that are so flexible as to describe anything and therefore have no substantive content. For example, our understanding of gamma-ray bursts has changed completely over the decades, but as a result of a flow of new data and a gradual evolution of competing ideas, each with many roots. Close examination shows no revolutions or paradigm shifts, if those terms have any meaning at all, but instead a gradual development consisting of a series of advances of greater or lesser importance. Progress has come from a process with many steps, most of them at least partly sideways, and with many contributors.

The idea of the genius scientist is as wrong, or as obsolete, as the idea of scientific revolution. Modern science is so competitive that ideas are brought forward half-formed and immediately seized on and further developed by someone else. They achieve their final perfected form, if they ever do, only after a long and complex evolution, drawing on many contributions and owing debts to many contributors. There are no modern Galileos, Newtons, or Einsteins (if even they were the lone genius creators of legend) because there are too many would-be Galileos, Newtons, and Einsteins. Just as in the stock market, when interest and enthusiasm are widespread, the prospect of reward is small.

Sometimes the philosophers speak plain nonsense: no scientist believes the truths of nature are figments of our assumptions. We cynically note that most philosophers of science are either failed scientists or ignorant of science. The most charitable conclusion may be that philosophy of science has nothing to do with science. Most scientists ignore it. Mathematics, at least, is useful.

We are usually more interested in the history of science. Perhaps this is because we hope to learn the secret of how to make great discoveries from the great scientists of the past. But there is no magic formula beyond a combination of talent, effort, insight, and luck. Perhaps the most important factor is the luck of being in the right place, geographically and intellectually, at the right time. We would like to understand how new scientific ideas are created, but this happens within the human mind, and leaves no paper trail for the historian. Even the discoverer himself generally does not understand how he arrived at his insight. For example, in *The Double Helix* James Watson describes the crucial step in understanding the structure of DNA: "Suddenly I became aware that. . . ." That's all. It has happened to every scientist, and although most of our insights are much more pedestrian, none of us understands where they come from.

Somewhat disappointed, we conclude that the history of science is chiefly of antiquarian and human interest. Scientists generally regard the efforts of professional historians of science as bogged down in minutiæ and irrelevancies while slighting the interesting scientific issues, and historians regard the occasional efforts of scientists to tell history as not up to their professional standards. Both are probably correct.

This book is a popular account, by a working scientist, of a field in which he was interested, and did research, over a quarter of a century. I have tried to get the facts right, both scientific and historical. It is not a work of historical scholarship, but I hope it may be useful as a source to anyone writing such a work.

Appendix. Did a Gamma-ray Burst Kill the Dinosaurs? Will a Burst Kill Us?

The gamma-ray bursts we observe are at enormous distances from us. When their redshifts have been measured, their distances turn out to be roughly ten billion light-years.* We observe them as weak sources of gamma rays or as faint sources of visible light, but it is natural to ask what would happen if a gamma-ray burst were much closer. They are rare events, few and far between, but occasionally one must occur within our galaxy, and perhaps in our neighborhood. Would it affect life on Earth? Did a gamma-ray burst kill the dinosaurs?

The last question is easiest to answer, and the answer is No. We can be confident that the dinosaurs were killed by the impact of a solid body like an asteroid or comet because the layer of rock laid down when they became extinct contains an unusually high abundance of iridium. This element is extremely rare in Earth's surface rocks, because it settled down into Earth's core along with most of Earth's iron. Iridium is comparatively common in meteorites because their iron has not settled out, and its presence is a telltale indicator of the impact of a large comet or asteroid. In addition, the iridium-rich layer of rock also contains tiny grains of shock-altered mineral, demonstrating that a large solid body hit Earth at that time.

Gamma-ray bursts will occasionally happen comparatively near Earth. Could one affect life? In one case, GRB990123, we know approximately how bright it was, in visible light, during the burst itself. Its redshift was measured from its afterglow to be 1.6. This means that, for perhaps 10 or 20 seconds, GRB990123 was a hundred thou-

---

*There may be one exception, GRB980425. If really associated with SN1998bw, it was about a hundred times closer.

sand times more luminous than our entire galaxy. If it had been at the distance of the Andromeda Galaxy (the closest large galaxy to our own) it would have been fifty times brighter than Venus. At the distance of the center of our galaxy (but in another direction, for there is an enormous amount of absorption of light by interstellar dust between us and the galactic center) it would have been more than a hundred times brighter than the full Moon. At the distance of the nearest star it would have been hundreds of thousands of times brighter than the Sun, like a nearby atomic fireball, killing and igniting all life on the side of Earth that was exposed to it.

Its gamma-ray intensity would have been about a thousand times greater still. Gamma rays are absorbed in the upper part of Earth's atmosphere, and don't reach the ground. With this much energy (the gamma-ray burst at the distance of the nearest star would deliver about as much energy per square centimeter as a 10-kiloton atomic bomb 6 feet away) the gamma rays heat, ionize, and tear off the uppermost 10% of Earth's atmosphere. There would be a powerful shock wave traveling downward, pulverizing anything not incinerated on the irradiated half of the Earth, and even carrying its destruction around to the other half, shielded from the burst's direct light and heat.

The energetic particles that make the burst carry much more energy than the visible light, or perhaps even the gamma rays, of the burst itself. However, these particles do not travel in straight lines, but instead wander around in interstellar space, their paths curved by the galaxy's magnetic field. They may even give up much of their energy pushing on the interstellar gas. They arrive gradually, spread over millions of years from a burst hundreds of light years away. As a result, their intensity (in this case, about ten times the present intensity of cosmic rays, and perhaps much less) would remain much too low to do any harm.

The visible light and gamma rays of a nearby gamma-ray burst can be so destructive because they travel in a straight line at the speed of light, delivering their energy in seconds. There would not be time for the air to carry the heat away. Even though gamma-ray bursts are about ten thousand times rarer* than supernovas, bursts may be a

---

*This number, and the estimate of one burst per million years per galaxy, assume that bursts radiate their energy equally in all directions. This is unlikely to be true, for their radiation is almost certainly strongly beamed. However, if the radiation is beamed, then the number of bursts must be greater by exactly the factor required to compensate for the fact that only a fraction of them shine their energy in our direction, and their effects on Earth are just as frequent and severe as if there were no beaming at all.

greater threat to life on Earth because their energy is delivered so rapidly in comparison to that of a supernova, which lasts weeks or months. In addition, supernovas make few gamma rays. The 10-kiloton bomb two arms' lengths away, delivering its energy in a few millionths of a second, would be much more destructive than the gamma-ray burst. But the bomb would affect only a tiny fraction of Earth's surface.

At the rate of one burst (shining in our direction) per million years per galaxy it is very unlikely that Earth was ever scorched by a gamma-ray burst at the distance of the nearest star, even allowing for Earth's age (4.65 billion years). The closest gamma-ray burst shining in our direction can be estimated (on statistical grounds, for there is no direct evidence) probably to have been a few hundred light-years away, about a hundred times more distant. Its visible light would have been several times brighter than the Sun. Radiant heat would have incinerated leaves, insects, and most small land plants and animals, and burned larger ones. There would have been widespread fires on half the Earth. A great extinction is possible, depending on how much of Earth's land area was on the side exposed to the burst, but there would be none of the telltale evidence (traces of iridium and grains of shock-altered rock) of a meteorite impact such as that which killed the dinosaurs. There is no proof such a gamma-ray-burst extinction happened in the past, but no good evidence against it either. Paleontologists recognize several great extinctions in the history of Earth, and the causes of some of them remain unknown.

Fortunately, there is something we can do about the threat of a nearby (by astronomical standards) gamma-ray burst. We cannot prevent two orbiting neutron stars from coalescing and making a burst—a puny spacecraft could have no effect on massive stars, even if we could send one that far. But if we know when and where a burst will occur, we can reduce the harm it will cause us, just as predicting the arrival of a hurricane gives us time to evacuate areas likely to be flooded.

If we can detect a neutron star binary and measure its orbit we can predict, to quite extraordinary accuracy, when the stars will coalesce. If it is detected a hundred years before coalescence, after a few decades of observation we will be able to predict the time of coalescence, and of the gamma-ray burst, to an accuracy of a few seconds. As the final climax approaches, the prediction will get even sharper.

If we predict a nearby gamma-ray burst to the second, then we can protect ourselves, just as we would from an advancing forest fire. Flammable brush may be removed from near buildings, and they may be wet with water or covered with fire-fighting foam. People can retreat indoors, or, if the burst is close and likely to be particularly intense, to underground shelters. For the closest bursts, people living on the hemisphere exposed to the burst could be evacuated to the side of Earth shielded from it.

The hardest task is to find the binary neutron stars before they coalesce. Because gamma-ray bursts are so intense, and so rare, we must look deep into space to find any potential threat. Old neutron stars are not easy to detect, especially at great distances, because they are very faint. The best hope may be to observe the faint X rays they emit as dilute interstellar gas trickles down onto them, or the dim glow of visible and ultraviolet light radiated by their hot surfaces as they slowly cool. This will not be easy—the power radiated may be less than the sunlight reflected by the planet Jupiter, and the distance ten million times greater!

With only one observed burst per million years per galaxy, the odds are that there are no threatening binary neutron stars close enough to endanger us. If we completely survey interstellar space out to a distance of a thousand light-years, and find no binary neutron stars which will coalesce in the next million years, we will know that we will be safe for that time. To look further into the future, we would need to look farther out into space, because a more distant binary neutron star might have time to approach us before it made a gamma-ray burst. However, knowing that we are safe for a million years would be enough to satisfy most people; it is not urgent to deal with the future beyond a million years!

The risk of a nearby gamma-ray burst is small, like the risk of a comet or asteroid hitting the Earth, but neither of these risks is quite zero. During the history of the Earth we know that an impact caused a great extinction and killed the dinosaurs, and a gamma-ray burst might be just as damaging. With some effort, this risk can be nearly eliminated.

Glossary

Accretion. Literally, increase, growth or adhesion, but in astronomy usually refers to a star or other mass that draws mass from its surroundings with its gravitational force. Accretion is particularly important for compact objects, such as neutron stars and black holes, because the gravitational binding energy released when they accrete matter is very large. Most astronomical sources of X rays are powered by accretion, as are quasars.

Angular momentum. A mechanical quantity that is conserved (does not change) in most orbits, specifically those in which force acts toward a fixed center of attraction. Because it is a fixed quantity (for the particular body in its orbit) it is a powerful tool for calculating the orbit. It has the form of mass × radius × velocity. Angular momentum is used throughout physics and mechanics, not just in astronomy.

Anisotropy. The property of not being the same in all directions, or of being unevenly distributed on the sky.

Arc-minute. $\frac{1}{60}$ degree (of angle), written '.

Arc-second. $\frac{1}{3600}$ degree (of angle), written ".

Asteroid. A small (the largest is 800 km across, but most are much smaller) rocky body. Most asteroids in our solar system have orbits between those of Mars and Jupiter, but a few go closer to the Sun. Very rarely, one collides with Earth.

Atomic number. The number of electrons in an atom, equal to the number of protons in its nucleus.

Background. Signal present in a detector even when no target is being observed. For example, cosmic rays and gamma rays from steady sources are background in a gamma-ray-burst detector.

Ball lightning. Rare form of lightning consisting of a glowing ball, typically about the size of a grapefruit, that hovers or moves slowly in the air. There is no satisfactory explanation of ball lightning, and there is some doubt it really exists, although there are numerous apparently credible eyewitness reports.

Baryon. A massive, strongly interacting elementary particle. The most familiar examples are protons and neutrons, which together make up the nuclei of atoms.

Baryon poisoning.   Reduction in the speed and Lorentz factor of pair fireball debris if it contains too many baryons, preventing it from making a gamma-ray burst.

BATSE.   Burst And Transient Source Experiment, a component of the Gamma-Ray Observatory. It made the crucial observations that established that gamma-ray bursts come from the far reaches of the universe.

BeppoSAX.   Italian–Dutch X-ray and gamma-ray-burst satellite launched in 1996 and expected to operate until 2002. It discovered the X-ray afterglows of gamma-ray bursts and provided the accurate positions that enabled ground-based astronomers to find their visible and radio afterglows.

Black body spectrum.   The characteristic spectrum, which depends on temperature, of electromagnetic radiation emitted by an object that absorbs (hence the term "black body") all radiation that falls on it. It is a reasonable approximation to the spectrum emitted by almost any body.

Black hole.   Intense gravitational field left behind by the collapse of matter to a point of infinite density. The point is not observable from outside, but its gravity is.

Brightness.   How bright an object appears to us. If it radiates isotropically, its brightness is its luminosity divided by 4π times the square of its distance from us ($\pi = 3.14159\ldots$).

CCD.   Charge-Coupled Device. An integrated circuit used to detect and record light in nearly all modern telescopes as well as in electronic cameras.

CGRO.   Compton Gamma-Ray Observatory, another name for GRO. NASA attaches the name of a famous scientist to its larger scientific satellites after their launch, so that Space Telescope became the Hubble Space Telescope and GRO became CGRO. Astronomers, used to the earlier name, often continue to use it.

Comet.   A small (typically 1–10 km across) icy body. In our solar system comets are in orbits that keep them far from the Sun nearly all the time. They become visible if they approach closer to the Sun because its radiation warms them, making some of the ice evaporate, and producing a tail of visible dust and vapor. Very rarely, one collides with Earth.

Compton scattering.   The scattering of electromagnetic radiation by electrons or positrons. It is significant chiefly for X rays and gamma rays.

Cosmic ray.   A very energetic particle, usually a proton, moving at nearly the speed of light. Most cosmic rays are trapped within our galaxy by its magnetic fields, but the most energetic also fill intergalactic space.

Counting statistics.   In a measurement consisting of counting discrete events (such as detections of gamma rays, or coin tosses) the result will include some random variation, so that if the measurement were repeated, a somewhat different result would be found. If N events are counted, the variation (called the standard deviation) is about $\sqrt{N}$, or a fraction $1/\sqrt{N}$ of the total. This limits the accuracy of the measurement, but the larger N is, the more accurate it is.

Cyclotron radiation.   Radiation emitted by charged particles moving in a magnetic field if their speeds are not close to the speed of light. It is produced by the same physical process as synchrotron radiation, but emitted by slower particles. Cyclotron radiation is only strong enough to be observed when the magnetic field is very large, as it is near some white dwarfs and neutron stars.

CZT. Cadmium zinc telluride, a material that can be used to make gamma-ray detectors with high energy resolution, but that operate at room temperature. This will permit future satellites to obtain good quality spectra of bursts and to answer definitively the question of whether their spectra contain lines.

Deconvolution. Removing the effects of blurring, smearing, or spreading (produced by an imperfect instrument, or passage of light through the unsteady atmosphere) of an image or spectrum. Always tricky, deconvolution may introduce artifacts, such as spurious spectral lines.

Doppler shift. If a source of radiation is moving with respect to an observer, the observer will measure it to have a different frequency and wavelength than if it were not moving. The difference is called the Doppler shift. The most familiar example is the variation in pitch of an approaching and receding siren. The expansion of the universe and the orbits of binary stars are inferred from their Doppler shifts.

Electromagnetic radiation. Energy carried by oscillating electric and magnetic fields. Familiar examples include radio waves, light (infrared, visible, and ultraviolet), X rays, and gamma rays. In vacuum it travels at the speed of light.

Electron. Negatively charged, very light, stable elementary particle that fills the outer parts of atoms.

Electron Volt (eV). The energy a single electron acquires when it travels freely between two surfaces whose electric potential differs by one volt. One eV equals 1.6×10^{-12} erg.

Entropy. A measure of disorder in a substance or signal. It is a thermodynamic quantity useful in understanding gas flows because it is usually conserved, just as angular momentum is useful in understanding orbits.

Erg. A (small) unit of energy, nearly equal to the energy required to lift a 1-milligram weight 1 cm in Earth's gravity. It is the unit of energy usually used in astronomy.

Error box. Region of the sky from which a signal may have arrived, or region in a graph in which some imperfectly measured quantity may be found. There are error circles, rectangles, annuli, odd-shaped polygons, and other shapes, depending on the instruments used to collect the data.

ETC. Explosive Transient Camera. An optical system designed to search for unpredictable flashes of visible light over a broad swath of sky.

Fluence. Cumulative flux, the total number of particles (or photons) that cross a unit area from beginning to end of some event (such as a gamma-ray burst).

Flux. Analogous to brightness or intensity, a flux is the number of particles (or photons) crossing a unit area per unit time.

Free parameters. Quantities in a theoretical model whose values are not known in advance, but which are chosen to fit the data. For example, the intrinsic pitch (frequency) and velocity of a moving siren may be free parameters I choose to explain a sound I hear. They are chosen to make the best fit between the model and the data. The values that give the best fit are likely to be close to its actual intrinsic pitch and speed. Most models of complex physical processes contain free parameters. If the model is known to be correct (I see an ambulance with flashing lights), then I can usually determine the values of the free parameters reliably. Unfortunately, if the model is wrong (e.g., the siren is not moving, and its pitch varies because its operator turns a knob), it may still fit the data well, but give entirely spurious values for the free parameters. The more of them there are, the easier it is for the model to fit the data, even if it is completely wrong. Hence, the fact that a model fits the data may not be sufficient to show that it is correct, especially if it has several free parameters.

Gamma ray. Electromagnetic radiation whose photons have energies greater than about 100,000 eV. Sometimes lower-energy photons (often as low as 10,000 eV) are also called gamma rays, overlapping the definition of X rays, especially if they are detected by instruments designed to detect higher-energy gamma rays.

Gauss. The usual unit of magnetic field. Earth's field is about 0.6 gauss. Laboratory magnets often have fields of about 10,000 gauss.

GRO. Gamma-Ray Observatory, a satellite launched 1991, deliberately destroyed 2000, which carried BATSE and other gamma-ray instruments.

GROCSE. Gamma-Ray-Burst Optical Counterpart Search Experiment, also known as GROCSE-I, based on the revived WFOVC at Livermore. It was used to search (unsuccessfully) for optical counterparts of gamma-ray bursts. Its planned successor, GROCSE-II, split into the LOTIS and ROTSE projects.

HETE-2. High-Energy Transient Explorer-2. A small satellite launched in 2000 to study gamma-ray bursts. Successor to the original HETE, whose launch failed in 1996.

Hypernova. A hypothetical subclass of supernova, suggested to be the source of gamma-ray bursts. SN1998bw was an unusually energetic supernova, and can reasonably be termed a hypernova, although its connection with gamma-ray bursts remains controversial. The term has also been suggested as a label for other hypothetical varieties of supernovas and novas.

Initial conditions. A very general term used to describe how something begins. It is believed the Big Bang was the initial condition of the universe. The initial conditions of a star are the distribution of density, temperature, and velocity in a cloud of gas that will become the star. Usually such initial conditions are poorly known. Some phenomena (such as stars) are believed to be nearly independent of their initial conditions, making it possible to calculate them. Others, particularly those that involve turbulent flows, may not be.

Inverse Compton scattering. Scattering of electromagnetic radiation by very energetic electrons or positrons. The radiation may be boosted to high energy, emerging as X rays or gamma rays.

IPN (InterPlanetary Network). A network of gamma-ray-burst detectors distributed around the solar system to determine the directions from which the bursts arise. Increasing the separation of the detectors increases the accuracy with which their directions can be determined. There have been several interplanetary networks, changing as new satellites are launched and old ones fail.

Isotropy. The property of being the same in all directions, or evenly distributed on the sky.

KeV. 1000 eV.

Light-year. 9.5×10^{17} cm, the distance light travels in a year.

Liouville's theorem. A mathematical result that limits the performance of optical systems. Physical processes that conserve entropy also satisfy Liouville's theorem.

Lorentz factor. The ratio of the total energy of a particle to its rest mass energy. In everyday life, Lorentz factors are a tiny bit greater than one (e.g., the Lorentz factor of a jet airplane in flight is about 1.0000000000005, with the 1 representing its rest mass energy and the 0.0000000000005 its kinetic energy, the energy of motion). For relativistic particles Lorentz factors may be very large (100–1000 for protons in a gamma-ray burst, and more than 10^{11} for the most energetic cosmic rays).

LOTIS. Livermore Optical Transient Imaging System, an instrument developed at Livermore as a successor to GROCSE to search for gamma-ray bursts in visible light.

Luminosity. The total power (energy per unit time) an object radiates.

MACHO. MAssive Compact Halo Objects. Hypothetical unseen dim stars contributing mass to the halo of our galaxy. Also, a telescope designed to search for them by observing gravitational microlensing. Compare to WIMPs, Weakly Interacting Massive Particles, another candidate for mass in a galactic halo.

Magnetar. A neutron star with magnetic field in the approximate range 10^{14}–10^{15} gauss, greater than the fields inferred for pulsars. There is good, but not completely conclusive, evidence that soft gamma repeaters are magnetars.

Magnitude. Astronomers' measure of brightness. It is a logarithmic scale, meaning that any two objects that differ by one magnitude differ by a factor of 2.512 in brightness. Larger magnitudes correspond to dimmer objects, so that a 20th magnitude star is 100,000,000 (twenty factors of 2.512 multiplied together) times fainter than a 0th magnitude star. The brightest star (other than the Sun) is −1st magnitude, the faintest seen by the naked eye is 6th magnitude, and the best telescopes can, with difficulty, observe stars as faint as about 26th magnitude.

MeV. 1,000,000 eV.

Microlensing. Amplification of the brightness of a distant object by the gravitational field of a mass between it and the observer. Used to search for the gravitational influence of otherwise unobservable dim masses.

Microwave background radiation. A relic of the Big Bang, microwave radiation filling the entire universe. It has the spectrum of a black body at a temperature of 2.7°K (absolute temperature scale).

Model. A mental picture or explanation of a complex phenomenon. Sometimes it is largely qualitative (a rotating magnetic neutron star is a model of a pulsar), and sometimes quantitative (a table of numbers describing the density, temperature, and pressure at each point in a star).

Neutron. Massive neutral particle found in atomic nuclei. On its own it is unstable, with a half-life of about 12 minutes, but it may be stable when bound in a nucleus.

Nova. An eruption on a white dwarf star, roughly a million times less energetic than a supernova. The star survives, and erupts again many times.

Order of magnitude. Used to describe any very rough estimate. Sometimes it refers specifically to an estimate to the nearest power of ten ("To order of magnitude, there are a billion people in China.").

Parallax. The effect that the motion of Earth produces small changes in the directions of stars. Used to measured the distances to stars.

Parsec. 3.1×10^{18} cm, the distance at which a star would have a parallax of 1 arc-second. A parsec is 206,265 times the mean distance between Earth and the Sun.

Peer review. The process by which scientific papers and grant proposals are reviewed by other scientists (the peers of those submitting them).

Photon. The fundamental unit of electromagnetic radiation, whose energy is proportional to the frequency of the radiation. Radiation has a character that is partly wave and partly particle; it is emitted and absorbed as discrete photons, with fixed amounts of energy.

Planck spectrum. Black-body spectrum, whose theory was developed by the physicist Max Planck in 1900.

Plasma. A gas whose particles are electrically charged, usually as a result of the removal of electrons from atoms. At a temperature of a few thousand degrees or higher, all matter becomes a plasma. Almost always, the numbers of positively and negatively charged particles are equal, so the plasma as a whole has no net charge.

Plasma wave. Waves in a plasma, often involving a small displacement of the positive charges from the negative charges.

Positron. The electron's anti-particle, with the same mass but positive charge. A positron and an electron, brought together, annihilate each other, releasing their rest mass energy as gamma rays.

Power law distribution. A distribution (of particle energies, for example) in which the number of particles is proportional to a power of the energy. More generally, the number could be proportional to a power of some other quantity.

Proton. Stable, massive, positively charged particle found in atomic nuclei.

Pulsar. A rotating magnetic neutron star. Like a lighthouse beacon, the intensity of its radiation (chiefly radio waves or X rays) is observed to vary at its rotation period. These periods range from about 0.001 second to about 5 seconds for radio pulsars, and as long as 1000 seconds for X-ray pulsars. The faster pulsars are powered by their rotational energy, and gradually slow. The slower pulsars (emitting only X rays) are powered by accretion of matter from a binary companion star, and their spin periods vary in a more complex and irregular manner. Known magnetic fields of pulsars are in the range 10^8–5×10^{13} gauss.

Quasar. Quasi-stellar object (QSO), a powerful (often hundreds of times more luminous than an ordinary galaxy) source of light, radio waves, and energetic particles. Quasars are believed to result from the accretion of gas onto a large black hole at the center of a galaxy. The galaxy itself may be undetectably faint, giving the quasar a star-like (point-like) appearance—hence the name. Quasars are "active galactic nuclei" (AGN), a more general term encompassing less luminous members of this family as well.

Rest mass energy. The energy mc^2 associated with a particle at rest of mass m, where c is the speed of light.

RMT. Rapidly Moving Telescope. A small telescope planned to follow up flashes of light observed by the ETC. It would have rapidly slewed to the patch of sky from which ETC observed the flash. It was designed to determine the location of the flash quite accurately (to about an arc-second), permitting identification and later close study by a large telescope.

ROTSE. Robotic Optical Transient Search Experiment, an instrument developed at Los Alamos as a successor to GROCSE to search for gamma-ray bursts in visible light. It succeeded on January 23, 1999.

Sigma. Standard deviation (because the Greek letter sigma is often used for standard deviation). A "three-sigma" result deviates from some prediction by three standard deviations, giving statistical evidence that the prediction was not correct. Hence, there may be some new phenomenon, not included in the prediction. Because of systematic errors, many scientists believe that about half of all three-sigma results are wrong, although if there were no systematic errors only about 0.3% of them would be.

Solar wind. A flow of dilute gas (typically 1–10 particles per cubic centimeter, not far above mean interstellar densities) away from the Sun at speeds of several hundreds of kilometers per second. It is powered by the release of solar magnetic energy.

Spectral line. Wavelength or energy at which an atom (or molecule, nucleus, particle, or electron) preferentially absorbs and emits radiation. They are called "lines" because of their appearance in a spectroscope.

Spectrum. The distribution of energy in electromagnetic radiation. It can refer to the distribution over broad ranges of wavelengths (e.g., blue vs. red light) or to narrow spectral lines.

Speed of light. 3×10^{10} cm/second, or 186,000 miles/second. The usual symbol is c.

Standard candle. A standard of luminosity, so named because an actual standardized candle was once the standard of luminosity. This term is used in astronomy to describe the assumption that all objects of a class have the same luminosity.

Standard deviation. The statistical uncertainty of a measurement. See **Counting statistics.**

Statistical fluctuations. Counting statistics.

Superluminal motion. Motion that appears to be faster than the speed of light, even though no object can move that fast. Superluminal motion may be observed, for example, if the spot of a laser is rapidly swept across a distant screen. The spot is not a physical object, and can move superluminally (the light in the spot at one time is not the same as the light in it at an earlier time, so the spot can move faster than the light itself).

Supernova. An explosion in which a massive star destroys itself. For some weeks supernovas are as bright as a typical galaxy. Supernovas may leave behind a neutron star, a black hole, or perhaps no compact object at all, in addition to the mass that is expelled in the explosion.

Supernova remnant. What is left over after a supernova, including an expanding shell of gas. If a pulsar was born, there are also magnetic fields and energetic particles created by the pulsar. The shell collides with the interstellar gas, emits radio waves and X rays, and is believed to accelerate cosmic rays.

Swift. Gamma-ray-burst satellite scheduled for launch in 2003.

Synchrotron radiation. Radiation emitted by a charged particle moving at nearly the speed of light in a magnetic field. Synchrotron radiation is generally significant only for electrons and positrons. First observed in particle accelerators called synchrotrons.

Systematic error. Error and uncertainty not caused by statistical fluctuations. Systematic errors are the consequence of not understanding an instrument perfectly, or of simply making a mistake. Because instruments are complicated and scientists human, systematic errors are ubiquitous and often impossible either to eliminate or to estimate.

Turbulence. Chaotic motion in a fluid, with many interacting waves or eddies.

Van Allen radiation belts. Regions of space near Earth (but above the atmosphere) in which many energetic particles are trapped by Earth's magnetic field. These particles are a significant hazard to and source of background for instruments in space.

Vela. (Spanish for watchman.) A system of U.S. satellites designed to monitor the Nuclear Test Ban Treaty. Twelve in all were launched, beginning in 1963. The final six, known as Advanced Vela, were launched 1967–70, and the system remained operational until 1985. Gamma-ray detectors on Advanced Vela satellites discovered gamma-ray bursts.

WFOVC. Wide Field-of-View Camera. Optical system built at Livermore for ballistic missile defense. Revived as GROCSE.

Whipple. A ground-based telescope for studying high-energy gamma rays.

White dwarf. A dim star about the size of Earth, but the mass of the Sun, which has exhausted nearly all its nuclear fuel. The Sun will become a white dwarf at the end of its life, in about 5 billion years.

X ray. Electromagnetic radiation with photon energies between about 100 eV and about 100,000 eV, although these limits are arbitrary and other values are often used. The definition of X rays may overlap the definition of gamma rays.

±. "Plus or minus." Used to indicate a range of uncertainty. For example, if your weight is 150 ± 5 pounds it is probably somewhere between 145 and 155 pounds. The uncertainty may be the result of measurement error or of real variations (you weigh more after you eat).

Sources

Scientific thought flows through several different channels with different degrees of permanence. One is in the minds of the scientists themselves. This is ephemeral and subject to the mutation of memory. I have used my own recollections, and have consulted numerous protagonists for their memories of events and ideas.

The second channel consists of contemporaneous written records. Experimentalists and observers generally keep systematic notebooks because it is essential to record the details and conditions of their experiments and observations, which cannot be recreated afterward. Theorists are not usually so systematic, because the foundations of a logical argument can be recreated. Computer codes should have this elaborate documentation, but rarely do; however, they played little part in the study of gamma-ray bursts. Laboratory notebooks are voluminous, sometimes barely legible, and not usually available to a historian unless he is concentrating on one or a very small number of individuals. I have had access to many of the notebooks of one major figure in this subject, but it would not have been feasible to examine those of all, even if they had been offered.

A third channel is the reminiscences of participants. Several wrote accounts of significant parts of the gamma-ray-burst story in response to my inquiry. I was amazed at how much effort they were willing to expend to ensure that their point of view was heard, and often found these very illuminating. This book is, to some extent, my own reminiscence.

The fourth channel consists of formal publications. Sometimes these appear in conference proceedings (instrumental designs are usually only published in these proceedings, if at all), and sometimes in professional journals. Since the mid-1990s most conference and journal papers in astronomy have also been published in electronic archives. This is chiefly a means of achieving overnight dissemination and more convenient access; the same papers are published both electronically and on paper.

Formal publications are carefully prepared, and considered documents of record, but they are not complete. They are supposed to present only new results, but different scientists interpret this differently. The background behind these results may be reported only sketchily, if at all. Sometimes, especially among theorists, a paper will simply stake a claim to ideas that are obvious or in general circulation, presenting little new insight. Such papers may be influential, especially if they are presented before a large audience at a conference and gain the author prominence. The prototypical example is the review paper, which does not purport to present new results, and which generally appears in conference proceedings or specialized review journals rather than the regular research journals. Some research papers have much of this character. Other papers may only narrowly report a new and original result or novel line of thought, leaving its broader implications implicit and unsaid. Often the spread of an idea is largely a matter of informal and oral arguments, poorly recorded in publications. For these reasons it may be quite difficult to trace its origins, especially in a vigorous and fast-moving field.

I have drawn on all these resources in writing this book. I have cited written reminiscences and mail, e-mail, and telephone conversations (all as "private communications"), in addition to formal publications. In some cases there has been good reason not to identify the sources of private communications. I have tried to credit the earliest independent appearance of an original or critical idea or result; sometimes there is more than one, arrived at independently and approximately simultaneously. However, the literature is so large, and so many ideas appear, apparently spontaneously, everywhere, that I have only been able to discuss and cite a few that seem to me the most important or interesting.

The reader seeking a technical discussion of gamma-ray bursts should consult one of the many professional review articles. At the time of writing the best is probably that by T. Piran published in *Physics Reports,* V. 314, 575–667 (1999). This will necessarily become obsolete in a few years, and many more will be written.

Preface

Units in science: Romer, R. H., *American Journal of Physics* V. 67, 13–16 (1999).

1: Vela

Discovery of gamma-ray bursts: Klebesadel, R. W., Strong, I. B., and Olson, R. A., *Astrophysical Journal Letters* V. 182, L85–88 (1973).

Popular account of discovery of gamma-ray bursts: Strong, I. B and Klebesadel, R. W., *Scientific American* V. 235, October 1976, 66–79A.

Vela satellites: *Encyclopedia Astronautica* (http://www.rocketry.com/mwade/craft/advdvela.html); Jet Propulsion Laboratory Mission and Spacecraft Library (http://msl.jpl.nasa.gov/QuickLooks/advelaQL.html).

2: Detectors

NaI(Tl) scintillators: O'Kelley, G. D., in *Methods of Experimental Physics* V. 5A *Nuclear Physics* (New York: Academic Press, 1961), 616–41.

3: Where Are They?

Ruderman's 1974 review talk: *Seventh Texas Symposium on Relativistic Astrophysics* (Proceedings of the New York Academy of Sciences V. 262), 164–80 (1975).

Usov, V. V., and Chibisov, G. V., *Soviet Astronomy-AJ* V. 19, 115–16 (1975).

Prilutski, O. F., and Usov, V. V., *Astrophysics and Space Science* V. 34, 395–401 (1975).

The classic paper on the sociology of assigning credit in science is entitled "The Matthew Effect" (Matthew 13:12 "For unto every one that hath shall be given, and he shall have abundance: but from him that hath not shall be taken away even that which he hath"): Merton, R. K., *Science* V. 159, 56–63 (1968).

4: What Are They?

Watson, J. D., *The Double Helix* (New York: Athenæum, 1968).

The model of gamma-ray bursts produced by comets falling onto neutron stars was first suggested by Harwit, M., and Salpeter, E. E., *Astrophysical Journal Letters* V. 186, L37–39 (1973).

The analogy to solar flares was suggested by Stecker, F., and Frost, K., *Nature Physical Science* V. 245, 70–71 (1973).

5: Compactness

The paradox of the Compton catastrophe was pointed out by Hoyle, F., Burbidge, G. R., and Sargent, W. L. W., *Nature* V. 209, 751–53 (1966).

It was resolved by Woltjer, L., *Astrophysical Journal* V. 146, 597–99 (1966).

The compactness problem was pointed out by Cavallo, G., and Rees, M. J. (*Monthly Notices of the Royal Astronomical Society* V. 183, 359–65 [1978]) and by Schmidt, W. K. H. (*Nature* V. 271, 525–27 [1978]).

Martin Rees: *Who's Who,* and private communications from Rees and others.

Couderc (1899–1981) published his work on light echoes in *Annales d'Astrophysique* V. 2, 271–302 (1939). A popular account is by Felten, J. E., *Sky & Telescope* V. 81, February 1991, 153–57. A biography is by de Vaucouleurs, G. and Walusinski, G., *l'Astronomie* V. 100, 409–15 (1986).

6: The Large Magellanic Cloud

Observations of the burst of March 5, 1979, were reviewed by Cline, T., *Comments on Astrophysics* V. 9, 13–21 (1980).

7: False Lines

Trümper, J., et al., *Astrophysical Journal Letters* V. 219, L105–10 (1978).

Mazets, E., et al., *Nature* V. 290, 378–82 (1981).

Murakami, T., et al., *Nature* V. 335, 234–35 (1988).

8: False Light

Schaefer's original results were published in Schaefer, B. E., *Nature* V. 294, 722–24 (1981) and Schaefer, B. E., et al., *Astrophysical Journal Letters* V. 286, L1–4 (1984).

Żytkow's criticism was published in Żytkow, A., *Astrophysical Journal* V. 359, 138–54 (1990).

Schaefer's reply was published in Schaefer, B. E. *Astrophysical Journal* V. 364, 590–600 (1990).

Anna Żytkow: http://www.his.com/~z/guestbook/zhurnal07.html; private communications, Żytkow and others.

"were never intended for the detection and analysis of fast optical transients": report of referee to Żytkow's 1990 paper.

Greiner, J., *Astronomy & Astrophysics* V. 264, 121–26 (1992).

Hudec, R., et al., *Astronomy & Astrophysics* V. 184, 839–52 (1994).

The unpublished correction to the position of the gamma-ray burst possibly associated with the supposed 1905 transient is cited by Gorosabel, J., and

Castro-Tirado, A. J., *Astronomy & Astrophysics* V. 337, 691–98 (1998) as K. Hurley, private communication (1997).

9: The Copernican Dilemma

"no tendency to cluster": Strong, I. B., Klebesadel, R. W., and Olson, R. A., *Astrophysical Journal Letters* V. 188, L1–3 (1974).

Fishman's first publication of balloon results was Fishman, G. J., et al., *Astrophysical Journal Letters* V. 223, L13–15 (1978).

The early Soviet results (obtained from spacecraft, not balloons) were published in Mazets, E. P., et al., *Astrophysics and Space Sciences* V.80, 1–143 (1981).

"Is your talk a joke?": V. V. Usov, private communication.

10: Soft Gamma Repeaters

The suggestion of the existence of a distinct class of repeating short bursts with anomalously soft spectra was apparently first made by Mazets, E. P., et al., *Astrophysics and Space Sciences* V. 84, 173–89 (1982). Further evidence was presented by Laros, J. G., et al., *Nature* V. 322, 152–52 (1986), and then in three nearly simultaneous papers: Atteia, J.-L., et al., *Astrophysical Journal Letters* V. 320, L105–10 (1987), Laros, J. G., et al., *Astrophysical Journal Letters* V. 320, L111–15 (1987) and Kouveliotou, C., et al., *Astrophysical Journal Letters* V. 322, L21–25 (1987).

The suggestion of an extraordinarily large magnetic field powering a burst by the release of magnetic energy was made by Katz, J. I., *Astrophysical Journal* V. 260, 371–85 (1982) as an explanation for the event of March 5, 1979, the first soft gamma repeater observed.

The term "magnetar" for a neutron star with a magnetic field in the range 10^{14}–10^{15} gauss, much greater than those found in radio pulsars, was suggested by Duncan, R. C., and Thompson, C., *Astrophysical Journal Letters* V. 392, L9–13 (1992) in connection with a model of classical gamma-ray bursts.

They developed magnetar models for soft gamma repeaters in a series of papers, beginning with Thompson, C., and Duncan, R. C., *Monthly Notices of the Royal Astronomical Society* V. 275, 255–300 (1995).

Colliding solid body models of soft gamma repeaters were discussed by Katz, J. I., Toole, H. A., and Unruh, S. H., *Astrophysical Journal* V. 437, 727–32 (1994).

Goodman, J., *Astrophysical Journal Letters* V. 308, L47–50 (1986).

Paczyński, B., *Astrophysical Journal Letters* V. 308, L43–46 (1986).

11: BATSE

Emergency spacewalk: J. Apt, private communication. Jay Apt is one of the two astronauts who did this spacewalk on Shuttle mission STS-37.

Mazets used satellites with six detectors to determine approximate burst positions (Mazets, E. P., and Golenetskii, S. V., *Astrophysics and Space Science* V. 75, 47–81 [1981]). Unfortunately, these detectors, piggybacked on planetary probes, were forty times smaller (in collecting area) and less sensitive than BATSE's, and could not detect the weak bursts necessary to test the N vs. S relation, or enough bursts to make a definitive measurement of anisotropy.

The initial publication of the BATSE results was by Meegan, C. A., et al., *Nature* V. 355, 143–45 (1992). There has been a long series of subsequent papers, mostly published in *Astrophysical Journal Supplement.*

Destruction of GRO and alternatives were discussed by Bortman, H., *Nature* V.405, 504–06 (2000), Park, R., *What's New*, June 2, 2000 (http://www. aps.org/WN/toc.html), and CBS News (http://cbsnews.cbs.com/now/story/ 0,1597,201777–412,00.shtml); private communications.

". . . safely returns to Earth": NASA press release June 5, 2000 (http://pao. gsfc.nasa.gov/gsfc/spacesci/structure/reentry.htm).

The truth was more devious: CBS News reported that NASA was concerned that if an uncontrolled re-entry of Mir harmed anyone the United States might be blamed. The full story was provided by a well-placed source, "Deep Space," who must remain anonymous.

12: The Great Debate

The original (1920) Great Debate is briefly discussed in many popular astronomy books, and has been the subject of much historical literature.

The gamma-ray burst debate was published in the December 1995 issue of the *Publications of the Astronomical Society of the Pacific* V. 107, 1131–76, including papers by the principals, a historical account of the 1920 debate by V. Trimble, and a review of gamma-ray bursts by G. J. Fishman.

"I believed in . . .": Trimble, V., *Publications of the Astronomical Society of the Pacific* V. 107, 1133–44 (1995).

"NSF's failure . . .": private communication.

XYZ affair: referee's report and private communications from X, Y, and Z.

Year after year . . . : NASA Astrophysical Theory Program made no awards for work on the theory of gamma-ray bursts 1994–97. After the discovery of afterglows answered many of the most important questions, one award was made in each of the years 1998, 1999, and 2000. Announcements of past awards are posted at http://spacescience.nasa.gov.

Lamb, D. Q., and Quashnock, J. M., *Monthly Notices of the Royal Astronomical Society* V. 265, L45–50, L59–64 (1993).

Rutledge, R. E., and Lewin, W. H. G., *Monthly Notices of the Royal Astronomical Society* V. 265, L51–56 (1993).

Narayan, R., and Piran, T., *Monthly Notices of the Royal Astronomical Society* V. 265, L65–68 (1993).

"Half of all three-sigma results are wrong." I have been unable to trace the origin of this scientific proverb, but I have been aware of it (and been counseled by it) since the early 1970s. Fortunately, if a scientific result is correct, improved measurement will soon prove it with much greater statistical significance.

"If it is essential to use probability to prove that you are right, you are usually wrong." Reported by Goudsmit, S. A., *Physical Review Letters* V. 10, 149–50 (1963).

13: The Theorists' Turn

The orbits and life expectancies of binary pulsars are discussed by Taylor, J. H., and Weisberg, J. M., *Astrophysical Journal* V. 345, 434–50 (1989).

The rate of neutron star coalescences in our galaxy is uncertain and controversial. One of the early papers is Phinney, E. S., *Astrophysical Journal Letters* V. 380, L17–21 (1991).

. . . killed blocking a car bomb: *New York Times*, October 30, 1998.

Shemi, A., and Piran, T., *Astrophysical Journal Letters* V. 365, L55–58 (1990).

Rees, M. J., and Mészáros, P., *Monthly Notices of the Royal Astronomical Society* V. 238, 41p–43p (1992).

Usov, V. V., *Nature* V. 357, 472–74 (1992).

Usov, V. V., *Monthly Notices of the Royal Astronomical Society* V. 267, 1035–38 (1994).

Fenimore, E. E., Madras, C. D., and Nayakshin, S., *Astrophysical Journal* V. 473, 998–1012 (1996).

Sari, R., and Piran, T., *Astrophysical Journal* V. 485, 270–73 (1997).

The roots of the arguments made by Fenimore, Madras, and Nayakshin and by Sari and Piran go back to Ruderman, M. A., *Seventh Texas Symposium on*

Relativistic Astrophysics (Annals of the New York Academy of Sciences V. 262), 164–80 (1975).

Figure 13-1 is modified from Katz, J. I., *Astrophysical Journal* V. 422, 248–59 (1994).

14: Afterglows

The subject of power law distributions in a wide variety of natural phenomena (mostly outside astronomy) has been popularized under the label "Self-Organized Criticality" (Bak, P., *How Nature Works: The Science of Self-Organized Criticality* [New York: Copernicus, 1996]), following development of a model and theory by Smalley, R. F., Turcotte, D. L. and Solla, S. A., *Journal of Geophysical Research* V. 90, 1894–1900 (1985), and Katz, J. I., *Journal of Geophysical Research* V. 91, 10412–20 (1986).

The $s = -\frac{1}{3}$ spectrum from radio through gamma-ray frequencies was predicted by Katz, J. I., *Astrophysical Journal Letters* V. 432, L107–9 (1994).

A study of BATSE gamma-ray data by Cohen, E., et al., *Astrophysical Journal* V. 488, 330–37 (1997) supported the predicted $s = -\frac{1}{3}$ spectrum. This conclusion was disputed by another study, also of BATSE gamma-ray data, by Preece, R. D., et al., *Astrophysical Journal Letters* V. 506, L23–26 (1998). The disagreement remains unresolved, probably because of the difficulty of deconvolving data from BATSE's NaI(Tl) detectors.

Afterglows from radio through X-ray frequencies, including radio self-absorption and the effects of electron energy loss, were predicted by Katz, J. I., *Astrophysical Journal* V. 422, 248–59 (Appendix) (1994). A radio afterglow was predicted on the basis of analogy to supernova remnants by Paczyński, B., and Rhoads, J. E., *Astrophysical Journal Letters* V. 418, L5–8 (1993).

BeppoSAX: http://www.asdc.asi.it/bepposax/.

The afterglow of the burst of February 28, 1997, was discovered by van Paradijs, J., et al., *Nature* V. 386, 686–89 (1997). Further details about this discovery, including an account of a controversy within the team over the allotment of credit, may be found in a biographical memoir by van Paradijs's longtime collaborator W. H. G. Lewin (http//xxx.lanl.gov/ps/astro-ph/0105344). This account was confirmed by an independent source.

A thermonuclear model of gamma-ray bursts, similar to accepted models of X-ray bursts, was published by Blaes, O., et al., *Astrophysical Journal* V. 363, 612–27 (1990).

The radio afterglow of the burst of May 8, 1997: Frail, D. A., et al., *Nature* V. 389, 261–63 (1997). This paper also confirmed the predicted $s = -\frac{1}{3}$ spectrum and self-absorption at lower radio frequencies.

$s = -0.44 \pm 0.07$: Galama, T. J., et al., *Astrophysical Journal Letters* V. 500, L97–100 (1998). Note that this paper uses a parameter α defined so that $\alpha = -s$.

$s = -0.25 \pm 0.04$: Frail, D. A., Waxman, E., and Kulkarni, S. R., *Astrophysical Journal* V. 537, 191–204 (2000). Note that this paper uses a parameter β defined so that $\beta = -s$.

The predicted (Figure 14-1) rise and fall of afterglow intensity at any given frequency were confirmed by Djorgovski, S. G., et al., *Nature* V. 387, 876–78 (1997).

The redshift (0.835) of the afterglow of the burst of May 8, 1997, was discovered by Metzger, M. R., et al., *Nature* V. 387, 878–80 (1997).

The redshift (3.42) of the host galaxy of the burst of December 14, 1997, was discovered by Kulkarni, S. R., et al., *Nature* V. 393, 35–39 (1998).

The redshift (4.50) of the afterglow of the burst of January 31, 2000, was discovered by Andersen, M. I., et al., *Astronomy & Astrophysics* V. 364, L54–61 (2000).

15: A Supernova Connection?

Quasars: Salpeter, E. E., *Astrophysical Journal* V. 140, 796–800 (1964).

Woosley, S. E., *Astrophysical Journal* V. 405, 273–77 (1993).

"Hypernova": Paczyński, B., *Gamma-Ray Bursts Fourth Huntsville Symposium* (New York: American Institute of Physics, 1998), 783–87.

The term "hypernova" was used for astronomical phenomena unrelated to gamma-ray bursts by Gass, H., Wehrse, R., and Liebert, J., *Astronomy & Astrophysics* V. 189, 194–98 (1988); Wilkins, P. N., and de Bruyn, A. G., *Monthly Notices of the Royal Astronomical Society* V. 242, 529–34 (1990); Park, S. J., and Vishniac, E. T., *Astrophysical Journal* V. 375, 565–67 (1991); and Colina, L., and Perez-Olea, D., *Monthly Notices of the Royal Astronomical Society* V. 259, 709–24 (1992).

GRB980425/SN1998bw: Kulkarni, S. R., et al., *Nature* V. 395, 663–69 (1998); Galama, T. J., et al., *Nature* V. 395, 670–72.

Statistical analyses of possible supernova/gamma-ray burst coincidences: Kippen, R. M., et al., *Astrophysical Journal Letters* V. 506, L27–30 (1998); Graziani, C., Lamb, D. Q. and Marion, G. H., *Astronomy & Astrophysics Supplement* V. 138, 469–70 (1999); http://xxx.lanl.gov/ps/astro-ph/9810374.

The entropy argument is unpublished.

baryon: *Oxford English Dictionary*.

Evidence for iron in gamma-ray bursts: Amati, L., et al., *Science* V. 290, 953–55 (2000); Piro, L., et al., *Science* V. 290, 955–58 (2000); Piro, L., et al., *Astrophysical Journal Letters* V. 514, L73–77 (1999).

16: The Holy Grail

ROTSE: C. Akerlof, private communication; http://www.umich.edu/~rotse/.

Discovery of simultaneous optical counterpart of GRB990123: Akerlof, C., et al., *Nature* V. 398, 400–402 (1999).

Boynton's design was published by Boynton, P., Kennicutt, R., and Tomandl, D., *Seventh Texas Symposium on Relativistic Astrophysics* (Annals of the New York Academy of Sciences V. 262), 209–13 (1975).

"failed to show . . ." P. Boynton, private communication.

ETC: Ricker, G. R., *High Energy Transients in Astrophysics* (New York: American Institute of Physics, 1984), 669–86; Vanderspek, R. K., Ricker, G. R., and Doty, J. P., *Robotic Telescopes in the 1990's* (San Francisco: Astronomical Society of the Pacific, 1992), 123–36; Vanderspek, R. K., Krimm, H. A., and Ricker, G. R., *Gamma-Ray Bursts Second Workshop* (New York: American Institute of Physics, 1994), 438–42; Krimm, H. A., Vanderspek, R. K. and Ricker, G. R., *Astronomy & Astrophysics Supplement* V. 120, No. 4, 251–54 (1996); Krimm, H. A., Vanderspek, R. K. and Ricker, G. R., *Gamma-Ray Bursts Third Huntsville Symposium* (New York: American Institute of Physics, 1996), 661–65; G. R. Ricker and S. D. Barthelmy, private communications.

RMT: Teegarden, B. J., et al., *High Energy Transients in Astrophysics* (New York: American Institute of Physics, 1984), 687–93 (1984); S. D. Barthelmy and G. R. Ricker, private communications.

A fortuitous failure . . . : *Sky & Telescope* V. 83, June 1992, 612; G. J. Fishman, private communication.

A new ground station . . . : J. Apt and G. J. Fishman, private communications.

BACODINE/GCN: Barthelmy, S. D., et al., *Gamma-Ray Bursts Second Workshop* (New York: American Institute of Physics, 1994), 643–47; *Gamma-Ray Bursts Third Huntsville Symposium* (New York: American Institute of Physics, 1996), 580–84; S. D. Barthelmy, private communication; http://gcn.gsfc.nasa.gov/.

GTOTE: *Bulletin of the American Astronomical Society* V. 29, 379–80 (1997); S. D. Barthelmy, private communication.

WFOVC: Park, H.-S., et al., *Proceedings of SPIE* V. 1111, 196–203 (1989); Park, H.-S., et al., *Proceedings of SPIE* V. 1304, 293–99 (1990).

"abandoned and unloved. . . .": C. Akerlof, private communication.

GROCSE: Akerlof, C., et al., *Gamma-Ray Bursts Second Workshop* (New York: American Institute of Physics, 1994), 633–37; C. Akerlof and H.-S. Park, private communications; http://www.phys.llnl.gov/V_Div/GROCSE/. GROCSE is also known as GROCSE-I.

"does not require a trigger. . . .": Krimm, H. A., Vanderspek, R. K., and Ricker, G. R., *Astronomy & Astrophysics Supplement* V. 120, No. 4, 251–54 (1996); Krimm, H. A., Vanderspek, R. K., and Ricker, G. R. *Gamma-Ray Bursts Third Huntsville Symposium* (New York: American Institute of Physics, 1996), 661–65.

GROCSE-II: C. Akerlof and B. C. Lee, private communications; see also ROTSE and LOTIS, which implemented the GROCSE-II design after the Livermore-Michigan effort split.

One prominent theorist . . . : I and others have observed this.

He withdrew . . . : J. Buckley, private communication.

LOTIS: H.-S. Park, private communication; http://hubcap.clemson.edu/~ggwilli/LOTIS/.

GROCSE data: Lee, B., et al., *Astrophysical Journal Letters* V. 482, L125–29 (1997); H.-S. Park, et al., *Astrophysical Journal* V. 490, 99–108 (1997).

Sari, R., and Piran, T.: http://xxx.lanl.gov/ps/astro-ph/9901105; *Astronomy & Astrophysics Supplement* V. 138, 537–38 (1999); *Astrophysical Journal* V. 520, 641–49 (1999).

Afterword

"Suddenly I became aware that. . . .": Watson, J. D., *The Double Helix* (New York: Athenæum, 1968), 194. Many other scientists, myself included, report such sudden inexplicable flashes of insight.

Appendix

The explanation of the extinction of the dinosaurs by the impact of an extraterrestrial body is now generally (but not universally) accepted, and is found in many textbooks. The original publication was by Alvarez, L. E., et al., *Science* V. 208, 1095–1108 (1980).

This appendix is based on my unpublished calculations.

A more alarming picture of the dangers of a comparatively nearby gamma-ray burst may be found in Leonard, P. J. T., and Bonnell, J. T., *Sky and Telescope* V. 95, 28–34 (February 1998) and in Dar, A., and DeRujula, A., http://xxx.lanl.gov/ps/astro-ph/0110162. I believe they have overstated its effects, particularly those of its cosmic rays, whose paths are bent by interstellar magnetic fields, and whose arrival is spread over a million years or more.

Index

Acceleration, of energetic particles, 13n, 180
  in particle accelerators, 43, 62
  in space, 132, 140–44, 152, 174
Acceleration, of fireball debris, 129–32, 134, 152, 159
Acceleration, shock, 142–43
Accretion, 35–36, 38, 70, 100–101, 103, 150–53, 189
Afterbursts, 94–95. *See also* Soft Gamma Repeaters, repetition
Afterglow. *See* Gamma-ray bursts, afterglows
Akerlof, Carl, 169–70, 172–73
Alligators, under beds, 124
Alpha rays (particles), 13
Andromeda nebula (galaxy), 117, 186
Angular momentum, 101–2, 113, 128, 158, 189
Archives, sky photographs, 72–80, 143
Asteroids, 100, 102, 178, 185, 188–89

Astrophysical Journal, 118–20
Astrophysics, theoretical, x, 104, 117–20, 161
Atmosphere, Earth's, 131, 186
Atoms, 139
Australia, 168

Background radiation, ix, 18, 70, 82, 164, 193
Backgrounds (in detectors), 4–5, 189
BACODINE, 168–71, 173–75, 178
Ball lightning, 79, 189
Balloons, scientific, 87, 89, 108
Barthelmy, Scott, 168–69
BATSE, 87, 92, 106–7, 113–15, 130, 179, 190. *See also* Gamma-Ray Observatory
  design, 108–9, 167–68
  distribution of bursts on sky, 109–11, 121–124
  implications, 111, 116–18, 126, 139
  intensities, 109–10

BATSE (*contiued*)
  localizations of bursts, 168–72,
    174–75, 178
  spectra of bursts, 143–44
BATSE Coordinates Distribution
    Network. *See* BACODINE
Baryon poisoning, 159–60, 163, 190
Baryons, 159, 189–90
Beaming of radiation, relativistic, 47–
    48
BeppoSAX, 145–47, 153–54, 175,
    190
Bessel, Friedrich, 22
Beta rays (particles), 13–15
Big Bang, 42, 45, 82
Black body spectrum, 190. *See also*
    Planck spectrum
Black holes, 70, 93, 111, 126–27,
    152, 158, 190
  accretion onto, 34–36, 102, 126,
    133, 150–53
  in gamma-ray bursts, 35, 100,
    104, 127–28, 133, 150, 152–53,
    158
Blueshift, 140
Bombs, nuclear. *See* Explosions,
    nuclear
Boynton, Paul, 165–66, 169
Bragg, Lawrence, 60
Bragg, William, 60
Brahe, Tycho, 22
Brilliant Pebbles, 169–70
Britain, 120
Burbidge, Geoffrey, 44–45

Cameras, 10, 162, 164, 170–71
  stereo, 165, 167
Cameras, X-ray. *See* Telescopes, X-
    ray
Canon, 162
Cavallo, Giacomo, 45–46
CCD. *See* Charge-Coupled Device
Cesium iodide. *See* Detectors,
    gamma ray, scintillation
CGRO. *See* Gamma-Ray Observatory
Chandra, 106
Charge-Coupled Device (CCD), 162,
    164–65, 171, 190
Chibisov, Gennadi, 89

Churchill, Winston, 39, 179–80
Coin, tossed, 5–7, 121–23
Comets, 26, 35, 100–102, 178, 190
  in extinctions, 185, 188
  in gamma-ray bursts, 30–33, 100
Compactness, 45–47, 49, 65–66, 94,
    104–5, 137–38
Compactness parameter, 47
Compton, Arthur, 44
Compton catastrophe, 44–46, 49
Compton Gamma-Ray Observatory
    (CGRO). *See* Gamma-Ray
    Observatory
Compton scattering, 15, 44–45, 129,
    190, 192
Conditions, initial, x, 31, 33, 38–39,
    159, 180, 192
Conservation
  of angular momentum, 101, 157–
    58
  of entropy, 157–58
Consistency, as a test of validity, 49,
    69, 80–81, 90
Copernican dilemma, 91–94, 99,
    108, 111
Copernican principle, 91
Copernicus, Nicolaus, 12, 82, 90
Cosmic rays, 1, 36, 131, 137, 140–
    41, 186, 190
Cosmic redshift. *See* Redshift,
    cosmological
Cosmological principle, 82, 86, 89
Cosmology, 23, 86, 106
Couderc, Paul, 48–49
Counting statistics, 5–7, 16–17, 56,
    67, 121–22, 190
Credit, assignment of in science, 28,
    49, 90, 172–73
Crick, Francis, ix, 29, 181
Currency, U.S., 155
Curtis, Heber, 116–17
Cyclotron radiation, 61–64, 190

Debate, The Distance Scale to
    Gamma-Ray Bursts (1995), 117–
    25
Debate, Great (1920), 116–17, 124–
    25
Deconvolution, 67, 69, 144, 191

Defects, photographic emulsion, 78
Detectors, gamma-ray, 2, 7, 9, 12–20, 23, 41, 108. *See also* Gamma-ray bursts, detectors
   angular resolution, 2, 7, 9, 16, 18–19
   CZT (Cadmium Zinc Telluride), 18, 191
   energy resolution, 16–18, 66–67, 69
   germanium (Ge), 17–18
   scintillation, 14–18, 61, 63, 66–67, 144
   Vela, 2, 7, 9
Diffraction gratings, 40–41, 60
Dinosaurs, 185, 188
Discovery, scientific, ix
Distances, astronomical, 21–23
DNA, 29, 181–82
Doppler shift, 42, 191
The Double Helix, 29, 182
Doughnut, neutronic, 128, 133–34, 136, 150, 152–53, 158
Dust grains, interstellar, 35

Earth, 185–87
Earthquakes, 79, 141
Echoes, light, 49, 135
Efficiency, conversion of mass to energy, 35–36, 126, 150
Ehrenfest, Paul, 122
Eighty Years' War, 12
Einstein, Albert, ix, 182
Einstein's cosmological constant, 106
Einstein's equation ($E = mc^2$), 35, 46, 52, 56, 126, 128, 130
Einstein's general theory of relativity, 127–28
El Alamein, 179–80
Electromagnetic radiation, 13, 40, 127–28, 133–34, 191
   wavelengths, 41
Electron-positron pairs, 15, 46–48, 56, 138
   annihilation, 56–57, 64–66, 104, 129
   gas, 66, 104, 152
Electron Volts (eV), 13

Electrons, 13, 15, 39, 62, 129, 131, 191
   high energy, 13n, 43–45, 132, 140–43, 174
   in electron-positron pairs, 46, 56, 129
Emulsions, photographic, 164
Energy
   accretional. *See* Energy, gravitational
   gravitational, 34–36, 100, 128
   kinetic, 34, 130, 132–33, 141
   magnetic, 34, 36, 38, 52, 58, 102
   orbital, 157
   rotational, 34, 36
   thermonuclear, 34–38
Entropy, 157–59, 191
Equilibrium, thermal, 66
Errors, in science, 81, 122
Errors, systematic, 122–24, 148, 196
Estimates, order of magnitude, 24
ETC. *See* Explosive Transient Camera
Europe, 120
Explosions, nuclear, 34–35, 37–39, 53, 186–87
   monitoring, 1, 3–5, 9–10, 16
Explosive Transient Camera (ETC), 166–72, 191
Extinction, 187–88

Federal Reserve Notes. *See* Currency, U.S.
Fees, investment management, 163
Fenimore, Edward, 134
Fermi, Enrico, 68, 141–42
Field of view, 73, 145–46, 164–67, 169–74, 178
Fire, forest, 187–88
Fireball, atomic, 186
Fireball debris, 129–37, 142, 144, 147, 150–52, 174, 180
Fireballs, 129–30, 132–35, 144, 150, 152, 159–60. *See also* Electron-positron pairs, gas.
Fishman, Jerry, 87, 92, 107–9, 114, 118
Fitting. *See* Parameter fitting
Flares, solar, 2, 36–39, 58, 102
Fluence, 85

Flux, 85
f-number, 162, 171
Focusing, camera. *See* Cameras
Focusing, gamma-rays
    (impossibility), 61
Focusing, X-rays, 60–61, 146
Free parameters, 68–69, 191–92
Friction, 131–32
Frost, Kenneth, 37
Funding, in science, 118, 165–66,
    172–73

Galactic disc, 25–27, 32, 51, 58
Galactic halo, 26, 91, 111
Galaxies, 116–17
Galaxies, radio, 164. *See also*
    Quasars
Galilei, Galileo, 12, 182
Gamma-ray burst Coordinate
    Network. *See* BACODINE
Gamma-ray bursts
  afterglows, 2, 144–50, 152–53,
    161, 175, 179–80
  beaming, 78, 137, 150–51, 160,
    180, 186
  brightness, 1, 33, 51, 55, 84
  collimation. *See* Gamma-ray
    bursts, beaming
  compactness problem, 45, 47, 49,
    65–66, 94, 104–5, 137–38
  coordinates. *See* Gamma-ray
    bursts, localization (directions
    to)
  detectors, 14–16, 61, 87, 108, 146
  difficulty of studying, 1–2
  discovery, ix, 1, 4–6, 40, 70, 164,
    178–79
  distances, cosmological, evidence
    against, 45–47, 66, 103–4, 120–
    21, 123–25, 139–40
  distances, cosmological, evidence
    for, 27, 88–92, 99, 112, 116, 123–
    25, 139, 148–50, 171, 179
  distances, cosmological,
    implications, 27, 34, 126–27,
    134, 136, 139, 150, 160–61
  distances, galactic, 26, 66, 69, 83,
    88, 90, 94, 98, 149–50, 179

distances, galactic halo, 91, 94,
    124–25
distances, measurement, 10, 23–
    27, 37, 82, 86–87, 148–50, 153,
    179, 185
distances, solar neighborhood, 26,
    32–34, 49, 77, 93, 99
distances, solar system, 26, 87
distribution on sky, 11, 26, 58, 83–
    84, 88, 94, 108
diversity, 7–8, 33–34, 37, 56, 58–
    59, 130, 133–34
duration, 1–2, 7–8, 37, 59, 74, 96,
    104, 126, 133
effects of nearby, 1, 185–88
energy, 1, 36–37, 76, 112–13, 126–
    27, 136, 150–51, 180
galactic hosts, 147–50
GRB670702 (July 2, 1967), 1, 4
GRB700822 (August 22, 1970), 5–
    6
GRB781119 (November 19, 1978),
    55–56, 74–75
GRB790107 (January 7, 1979), 94–
    95. *See also* Soft Gamma
    Repeaters
GRB790113 (January 13, 1979),
    75
GRB790305 (March 5, 1979),
    20, 51–52, 55–59, 62, 64, 83,
    91, 94, 96, 98–99, 102–3,
    155. *See also* Soft Gamma
    Repeaters
GRB791105 (November 5, 1979),
    75
GRB970228 (February 28, 1997),
    147
GRB970508 (May 8, 1997), 147–
    48
GRB971214 (December 14, 1997),
    150
GRB980425 (April 25, 1998), 153–
    56, 160, 185n
GRB990123 (January 23, 1999),
    162, 175–77, 185–86
GRB000131 (January 31, 2000),
    150
heterogeneity, 7–8

identification with other astronomical objects, 11, 38–39, 52–53, 59, 83
isotropy, 26–27, 32, 83–84, 89–91, 109–12, 121–24
localization (directions to), 179
localization by detector networks, 9–11, 19, 52, 179
localization by position-sensitive detectors, 109, 146, 168, 175, 179
localization, difficulty, 2, 93, 157, 163
localization, implications, 59, 161
localization of visible transients, 166–68
luminosity, 45, 104
models, 30–39, 126–38, 171, 179–80
models of components, 76–78, 92–94
models of energy source, 30–36, 126–29, 133–34, 150–53, 159–60
models of radiation processes, 64, 104–5, 130–32, 137–38, 143–44
models of time structure, 54, 104–5, 132–37
number-flux relation, 87–91, 94, 108–10
power, 36, 113, 126
radio counterparts, 143–45, 154
rate, 27, 96, 98, 160, 186–87
repetition, 74–78, 98, 123–24
subpulses, 5–8, 34, 37–38, 133–37
spectra, continuum, 96, 99, 104–5, 126, 137, 140, 143–44
spectra, line, 60–69, 90, 94, 103–4, 139–40, 161, 179
supernova counterparts, 153–54, 156–57, 160
time-dependence, 5–8, 33–34, 37, 51–54, 58, 94, 104, 133–37, 151
visible counterparts (archival), 74–81, 94, 103–4, 124, 143, 166, 179
visible counterparts (non-simultaneous), 19–20, 70–81, 93, 108–9, 140, 147, 154, 179
see also Gamma-ray bursts, afterglows)
visible counterparts (simultaneous), 143–45, 162–78
X-ray counterparts, 154, 161
Gamma-Ray Observatory (GRO), 87, 106, 108, 178, 190, 192
destruction, 113–15
tape recorders, 167–68
Gamma-ray Optical Counterpart Search Experiment (GROCSE), 170–73, 192
Gamma-ray sources, steady, 26, 100, 170
Gamma-rays, properties, 4, 7, 12–15, 41, 60, 66, 192
Gamma-ray To Optical Transient Experiment (GTOTE), 169–70
GCN. *See* BACODINE
Geiger counter, 14
General theory of relativity, 127–28
Globular clusters, 26
Goldin, Dan, 114
Goodman, Jeremy, 104, 126, 129–30
Grants, to support research, 118–20, 170
Gravitational radiation (waves), 127–29
Great Observatories, 106–7, 115
Greiner, Jochen, 80
GRO. *See* Gamma-Ray Observatory
GROCSE, 170–73, 192
GROCSE-II, 172–73
GTOTE, 169–70
Guam, 168
Gutenberg-Richter law, 141
Gyroscopes, 113–14

Halo, galactic. *See* Galactic halo
Harriot, Thomas, 12
Harvard College Observatory, 72
Harwit, Martin, 30, 100
Helium, 158
Hercules X-1, 62–64, 77
HETE, 178, 192
HETE-2, 178, 192

High Energy Transient Explorer-2
(HETE-2), 178, 192
Historians of science, 182–83
Holland, 12
Hoyle, Fred, 44–45
Hubble, Edwin, 42, 117
Hubble Space Telescope, 106, 114,
147, 161
Hudec, René, 80
Hurricane, 187
Hydrogen, 61, 129, 158
Hypernovas, 153, 157–61, 192
Hypotheses, scientific, 124

Ill-posed problems, 66–67
Image intensifier, 171
Income distribution, 141
Initial conditions. *See* Conditions,
initial
Interplanetary Networks (IPN), 179,
192
principles, 9, 19–20, 108–9, 163,
175, 192
results, 23, 71–72, 75, 83, 179
Interstellar clouds, 26, 133, 135–37,
141
Interstellar dust, 161, 164–65, 186
Interstellar gas, 130–31, 186, 188
and burst afterglows, 144, 148
and gamma-ray bursts, 131–33,
142, 174, 180
heterogeneity, 132–33
Interstellar magnetic field, 186
Interstellar space, 141, 186, 188
Inverse Compton scattering, 44n,
192. *See also* Compton
scattering
Inverse square law, 25, 84–85
Ions, 15, 139
Iridium, 185
Iron, 161
Iron, spectrum, 61, 148, 161
Israel, 120, 128

Jets, astronomical, 151–52, 160
Jupiter, 100, 146, 156, 188

Keck telescope, 148, 150
Kopernik, Mikołay. *See* Copernicus,
Nicolaus

Lamb, Donald, 120–25
Large Magellanic Cloud, 51–52, 98,
103
Lawrence Livermore National
Laboratory, 169–73
Laws of nature. *See* Laws, physical
Laws, physical, x, 81
Lee, Brian, 170, 173
Lenses, 162–66
Lewin, Walter, 123
Light echoes, 49, 135
Light, visible, properties, 13, 41
Lightning, 48
Linux, 162
Liouville's theorem, 164–65, 171,
193
Lippershey, Hans, 12
Livermore Optical Transient
Imaging System (LOTIS), 173–
75, 177–78, 193
LMC. *See* Large Magellanic Cloud
Lorentz factor, 130, 132, 137–39,
141–42, 144, 163, 193
Los Alamos (National Laboratory),
120, 162, 169, 173
LOTIS, 173–75, 177–78, 193

MACHO, 170, 173, 193
de Magalhães, Fernão. *See* Magellan,
Ferdinand
Magellan, Ferdinand, 50–51
Magellanic Clouds, 51
Magnetars, 38, 103, 133–34, 193
Magnetic fields, 33–34, 36, 57, 62–
63, 132, 134, 141
of neutron stars, 38, 52, 54–55,
58, 62–64, 66, 102
of stars, 38, 70
of Sun, 38
Magnitude scale, 73
Mars, 100
Masks, shadow, 146
Mathematics, 181–82
Mazets, Evgeni, 64, 87, 90, 92, 109
Mészáros, Peter, 130, 133–34
Meteorites, 80, 185
Meteor showers, 102
Microlensing, gravitational, 120,
170, 193

Microwave background radiation, ix,
18, 70, 82, 164, 193
Milky Way, 23, 25, 51, 116. *See also*
Galactic disc
Mir, 114
Missing mass, 111, 180
Models, 193
astrophysical, 101, 134
lighthouse, 54, 63
of gamma-ray bursts. *See* Gamma-
ray bursts, models
of soft gamma repeaters. *See* Soft
Gamma Repeaters, models
of Sun, 29
Moon, 100–101, 107, 145, 186
M87, 151

Narayan, Ramesh, 124
NASA. *See* National Aeronautics and
Space Administration
National Academy of Sciences, U.S.,
116, 120
National Aeronautics and Space
Administration (NASA), 108,
113–15, 168
Goddard Space Flight Center, 166,
168
Great Observatories, 106–7
Marshall Space Flight Center, 87
research funding, 18, 118, 120,
169, 172
National Science Foundation (NSF),
118, 120, 165–67, 172
Nebulas 116–17. *See also* Galaxies
Neighborhood, solar, 26, 32, 34, 47
Neutrinos, 113n, 127–28, 159
Solar neutrino problem, 180
Neutrons, 128, 193
Neutron stars, 35, 53–54, 70, 111–
13, 133. *See also* Magnetars,
Pulsars
accretion onto, 30–36, 53, 100–
102, 147, 188
binary, 57, 76–78, 93, 127–28,
147, 187–88
formation, 127–28, 158, 161
coalescence, 127–28, 153, 159–61,
187–88
collision, 104, 127

entropy, 157–59
gamma-ray burst models, 35, 39,
56, 104, 113, 126–28, 133, 136
gravity, 56, 159
magnetic field, 36, 53–55, 57–58,
62, 102–3, 133–34, 159
rotation, 34, 36, 54–55, 57, 62,
102–3, 128, 133
Newtonian mechanics, 81. *See also*
Angular momentum
Newton, Isaac, 182
Notation, scientific, x
Novas, 34, 37, 48–49, 193
Nozzles, 151
NSF. *See* National Science
Foundation
Nuclear Test Ban Treaty, 1, 3, 181
Nuclei, atomic, 141, 159
Number-Flux relations, 85–86

Ockham's razor, 25

Paczyński, Bohdan, 104, 120, 125–
26, 129–30, 153
Pairs. *See* Electron-positron pairs
Paradigms, 182
Parallax, 22–23, 194
Parameter fitting, 67–69, 151, 180
Park, Hye-Sook, 169–70, 172–73
Particles, energetic, 1, 36, 137, 141,
186. *See also* Electrons, Protons
Pearl Harbor, Hawaii, attack on, 3
Peer review. *See* Reviewing
Philosophers of science, 181–82
Photographs, astronomical, 79n
Photomultiplier tube, 15–17, 67
Photons, 13, 164, 194
Physics, compared to astrophysics,
31
Picture elements (pixels), 162, 164–
65, 171
Pillows, 159
Piran, Tsvi, 124, 128–30, 132, 134,
174–75
Pixels. *See* Picture elements
Planck, Max, 66
Planck spectrum, 66, 104, 126, 194
Planets, 35, 100–101, 134
colliding, 101–2

Plasmas, 39, 129, 131, 134, 137, 157, 194
  waves, 131, 194
Positrons, 13, 46, 56, 62, 194
Portugal, 50
Power law distributions, 141–42, 149, 194
Prediction, contrast to explanation, 144
Prisms, 40–41, 60
Progress, scientific, ix
Protons, 13, 129, 131–32, 141–42, 159, 194
Proverbs, scientific, 90, 122
Pulsars, 31, 57–58, 156, 180, 194
  binary, 127
  birth, 31, 54, 127
  discovery, ix, 62, 164
  distribution in space, 26, 111–12, 122
  magnetic fields, 54–55, 62, 103, 133–34
  radio, 34, 58
  rotation, 54, 57, 62, 102–3, 133–34
  X-ray, 57–58, 62

Quasars, 40, 104, 151, 156, 180, 194
  discovery, ix, 40, 45, 70, 93, 140, 152
  distances, 24–25, 42, 45, 85–86
  models, 42–48, 120
Quashnock, Jean, 120–24

Radio sources, astronomical, 40, 85–86
Radio waves, properties, 13
Rapidly Moving Telescope (RMT), 166–69, 172, 195
Real time, 166–68
Redshift, cosmological, 42, 70, 85, 139–40, 153
  gamma-ray bursts, 112, 139–40, 148–50, 153
  quasars, 42, 45, 85–86
Redshift, gravitational, 56–57
Rees, Martin, 45–46, 48–49, 130, 133–34
Refereeing. *See* Reviewing
Refsdahl, Sjur, 120

Relativity, general theory, 127–28
Relativity, special theory, 48, 81, 135–36
Replication, as a test of validity, 78, 80–81, 123
Research Corporation, 172
Reviewing, 194
  grant proposals, 118, 165–66, 172
  papers, 118–19
Ricker, George, 166
Risk reduction (of nearby gamma-ray burst), 187–88
RMT, 166–69, 172, 195
Robotic Optical Transient Search Experiment (ROTSE), 162, 164, 173–78, 195
ROTSE, 162, 164, 173–78, 195
ROTSE-II, 178
ROTSE-III, 178
Ruderman, Malvin, 27, 33, 36
Russia, 39, 114
Rutledge, Robert, 123

s. See Spectral index
Salpeter, Edwin, 30, 100, 152
Sargent, Wal, 44
Sari, Re'em, 134, 174–75
Satan, 156
Schaefer, Bradley, 74–80, 166
Schmidt, Maarten, 42
Schmidt, Wolfgang, 45–46
Science
  big, 170, 172
  debates, 124
  experimental, 181
  fashions in, 92
  geniuses in, 182
  history of, 182–83
  models in, 29–30, 38, 68, 178, 193
  philosophy of, 180–82
  progress of, ix, 178–82
  revolutions in, 182
  role of technology, 12, 163–65, 167–68, 171, 181
  Soviet, 90
  theoretical, 90, 181
Scintillation, 15
  radio, 147
  starlight, 147

Scintillator. *See* Detectors, gamma-ray, scintillation
Sea urchins, 47
Self-absorption, 144, 147–49
SGR. *See* Soft Gamma Repeaters
Shapley, Harlow, 116–17
Shemi, Amotz, 128–30, 132
Shocks, 136–37, 141–43, 158, 174, 180, 185–86
    external, 135–37, 143–44
    internal, 135–37, 143–44
    nonrelativistic, 142–43
    relativistic, 142–44, 148, 174
Sigma. *See* Standard deviation
Skylab, 113
Slewing, 166, 169
Small Magellanic Cloud, 51
Smart Rocks, 169
SMC, 51
Soda straw, 71
Sodium iodide. *See* Detectors, gamma-ray, scintillation
Soft Gamma Repeaters, 94–103, 150
    duration, 95–97
    models, 99–103
    repetition, 94–96, 102, 124
    spectra, 52, 55–59, 96, 98
Solar system, 23–24, 26
Solomon, 173
Sonic boom, 137
Spain, 12, 50
Space Shuttle, 107
Space Station, 107, 113–14
Spectra, 41, 142–43, 195
Spectral index (*s*), 142–44, 148–49, 174
Spectral lines, 41–42, 64, 70, 77, 139–40, 148, 195
    absorption, 64–65, 68
    annihilation, 56–57, 64–65, 139–40
    cyclotron, 61–66, 90, 139
    Doppler shifts, 64–66, 139
    emission, 61–65, 68
    gamma-ray, 61–67, 140
    hydrogen, 140
    iron, 61, 148, 161
    magnesium, 140, 148
    X-ray, 61–63

Spectroscopes, 40–41
Spectroscopy, 60, 140
Speed, of optical system (f-number), 162, 171
Spice Islands, 50
Standard candles, 21–23, 117, 195
Standard deviation, 121–22, 195. *See also* Counting statistics
Stars
    binary, 63, 70, 77, 127, 147, 163
    Cepheid, 117
    collapse, 127–28, 158
    composition, 61, 70
    death, 131, 161
    exploding, 2, 127, 158. *See also* Supernovas
    flare, 38
    formation, 161
    massive, 26, 158, 161
    naked-eye, 26, 58, 74, 163
    neutron. *See* Neutron stars
    planets, 101
    pulsating, 70
    rotation, 70, 158
    spectroscopy, 60, 70
    spots, 70
    structure, 158
    supergiant, 158
    variable, 2, 70–74, 156, 178. *See also* Stars, Cepheid
    white dwarf, 35, 39, 48, 196
    X-ray. *See* X-ray sources.
Star Wars, 169, 171
States, excited, 15
Statistical fluctuations. *See* Counting statistics
Statistical inference, 121–22, 155–56. *See also* Counting statistics
    a posteriori, 155–56
    a priori, 156
Statistical significance, 121–22, 155
Stecker, Floyd, 37
Stock market, 182
Strategic Defense Initiative (SDI), 169, 171
Strings, cosmic, 35
Sun, 146–47, 158, 186–87
Superbursts, 94, 96, 98, 100, 102–3. *See also* Soft Gamma Repeaters

Super-LOTIS, 178
Superluminal (faster than light)
expansion, 48–49, 195
Supernova remnants, 26, 141–42,
144, 195–96
pulsars, 127
soft gamma repeaters, 52–53, 55,
58, 94, 101–2, 155, 179
Supernovas, 34, 186–87, 195
and gamma-ray bursts, 153–61
and soft gamma repeaters, 101–2
distance measures, 22–23
in stellar evolution, 31, 37, 40,
113, 127
SN1572, 22
SN1987A, 49, 51, 113, 127
SN1998bw, 153–54, 156–57, 160–
61, 185
Swift, 178, 196
Synchrotron radiation, 43–46, 62,
132, 137, 139–44, 196

Taxes, 163
TDRS. *See* Tracking and Data Relay
Satellites
Teegarden, Bonnard, 166
Telescopes, 16, 40, 71–73, 90, 114–
15, 147, 164, 178. *See also*
Hubble Space Telescope
robotic, 162, 165–66, 172
X-ray, 60, 106, 145–47
Thallium. *See* Detectors, gamma-ray,
scintillation
Thermal equilibrium. *See*
Equilibrium, thermal
Thought experiments, 65
Tracking and Data Relay Satellites
(TDRS), 166–68
Trümper, Joachim, 63
Turbulence, 31–33, 38–39, 153, 160,
196

magnetohydrodynamic, 134, 152
plasma, 39, 131–32, 137, 152

Units, x
Universe, expansion, 42, 117, 164
University of Michigan, 172–73
Usov, Vladimir, 27–28, 49, 89–90,
92, 112, 133–34, 136

Van Allen radiation belts, 146, 196
van Maanen, Adriaan, 117
van Paradijs, Jan, 147
Vela satellites, 1, 3–4, 9–11, 18–19,
102, 196
discovery of gamma-ray bursts, 1,
4–11, 178
implications of data, 10–11, 23,
86–87, 108–9, 163
Venus, 186
Vidicon, 165

Watson, James, ix, 29, 90, 181–82
WFOVC, 169–73, 196
Whipple, 170, 173, 196
White holes, 35
Wide field-of-view camera
(WFOVC), 169–73, 196
William of Ockham, 25
Windows (operating system), 162
Winds, stellar, 70, 133, 158, 161,
195
Woltjer, Ludwig, 45
Woosley, Stanford, 153, 158
Wormholes, 35

X-ray bursts, 2, 147, 163
X-ray sources, ix, 26, 31, 34, 100,
163, 188
X-rays, properties, 4, 13–14, 60, 196
XYZ Affair, 118–19

Żytkow, Anna, 79–80
Zwicky, Fritz, 111